Endangered Species - Present Status

Edited by Mohammad Manjur Shah

Published in London, United Kingdom

Endangered Species - Present Status
http://dx.doi.org/10.5772/intechopen.107790
Edited by Mohammad Manjur Shah

Contributors
Qing Ma, Gang Wu, Wenjie Li, Seyit Yuzuak, Fachun Guan, Yin Lu, Patricia Yu, Sanjay Saraf, Marjorie Riofrío-Lazo, Diego Páez-Rosas, Suman Acharya, Tadeu Gomes de Oliveira, Fábio Dias Mazim, Paulo Guilherme Carniel Wagner, Alisson da Rosa Boÿink, Lester Alexander Fox-Rosales

First published in London, United Kingdom, 2023 by IntechOpen
IntechOpen is the global imprint of INTECHOPEN LIMITED, registered in England and Wales, registration number: 11086078, 5 Princes Gate Court, London, SW7 2QJ, United Kingdom

British Library Cataloguing-in-Publication Data
A catalogue record for this book is available from the British Library

Additional hard and PDF copies can be obtained from orders@intechopen.com

Endangered Species - Present Status
Edited by Mohammad Manjur Shah
p. cm.

This title is part of the Environmental Sciences Book Series, Volume 13
Topic: Ecosystems and Biodiversity
Series Editor: J. Kevin Summers
Topic Editors: Salustiano Mato, Josefina Garrido and Francisco Ramil

Print ISBN 978-1-83769-060-2
Online ISBN 978-1-83769-061-9
eBook (PDF) ISBN 978-1-83769-062-6
ISSN 2754-6713

For EU product safety concerns:
IN TECH d.o.o., Prolaz Marije Krucifikse Kozulić 3, 51000 Rijeka, Croatia,
info@intechopen.com or visit our website at intechopen.com.

IntechOpen Book Series

Environmental Sciences

Volume 13

Aims and Scope of the Series

Scientists have long researched to understand the environment and man's place in it. The search for this knowledge grows in importance as rapid increases in population and economic development intensify humans' stresses on ecosystems. Fortunately, rapid increases in multiple scientific areas are advancing our understanding of environmental sciences. Breakthroughs in computing, molecular biology, ecology, and sustainability science are enhancing our ability to utilize environmental sciences to address real-world problems.

The four topics of this book series - Pollution; Environmental Resilience and Management; Ecosystems and Biodiversity; and Water Science - will address important areas of advancement in the environmental sciences. They will represent an excellent initial grouping of published works on these critical topics.

Meet the Series Editor

J. Kevin Summers is a Senior Research Ecologist at the Environmental Protection Agency's (EPA) Gulf Ecosystem Measurement and Modeling Division. He is currently working with colleagues in the Sustainable and Healthy Communities Program to develop an index of community resilience to natural hazards, an index of human well-being that can be linked to changes in the ecosystem, social and economic services, and a community sustainability tool for communities with populations under 40,000. He leads research efforts for indicator and indices development. Dr. Summers is a systems ecologist and began his career at the EPA in 1989 and has worked in various programs and capacities. This includes leading the National Coastal Assessment in collaboration with the Office of Water which culminated in the award-winning National Coastal Condition Report series (four volumes between 2001 and 2012), and which integrates water quality, sediment quality, habitat, and biological data to assess the ecosystem condition of the United States estuaries. He was acting National Program Director for Ecology for the EPA between 2004 and 2006. He has authored approximately 150 peer-reviewed journal articles, book chapters, and reports and has received many awards for technical accomplishments from the EPA and from outside of the agency. Dr. Summers holds a BA in Zoology and Psychology, an MA in Ecology, and Ph.D. in Systems Ecology/Biology.

Meet the Volume Editor

Dr. Mohammad Manjur Shah obtained his Ph.D. from Aligarh Muslim University, India, in 2003. He is a pioneer in the field of insect parasitic nematodes and has presented his findings at several conferences, published numerous journal articles, and edited eight books. He is also a journal reviewer. Dr. Shah completed two postdoctoral fellowships under the Ministry of Science and Technology, Government of India, before joining Yusuf Maitama Sule University, Nigeria, in 2015. He is currently an Associate Professor of Biology, Director of Research Innovation and Development, and Chairman of the Research Ethics Committee at Yusuf Maitama Sule University.

Contents

Preface

According to the International Union of Conservation of Nature (IUCN), endangered species are those categorized as very likely to become extinct in their own native ranges in the near future. Per various reports, thousands of species are lost every year, and their numbers continue to increase annually at an alarming rate. As such, this book provides a comprehensive report of endangered animal and plant species. Some of the endangered animal species discussed include Chinese pangolins, Pompa cats, and Galapagos pinnipeds (sea lions and fur seals). Endangered plant species examined include dragon blood trees and plants native to Hawaii. I hope readers will find this book a useful resource for enhancing their knowledge of endangered species.

Dr. Mohammad Manjur Shah
Associate Professor,
Department of Biological Sciences,
Yusuf Maitama Sule University,
Kano, Nigeria

Section 1

Conservation of Endangered Animal Species

Chapter 1

Galapagos Pinnipeds, Challenges to Their Survival

Marjorie Riofrío-Lazo and Diego Páez-Rosas

Abstract

Pinnipeds endemic to the Galapagos archipelago are in endangered conservation status. The Galapagos sea lion, *Zalophus wollebaeki*, and Galapagos fur seal, *Arctocephalus galapagoensis*, have adapted to an ecosystem with high environmental variability and unpredictable marine productivity to survive. In addition to the environmental factors that pressure their populations, these species are exposed to anthropogenic influence, mainly in rookeries on islands with human settlements. It has been determined that the populations of Galapagos pinnipeds have different growth trends between regions of the archipelago, islands of the same region, and between rookeries of the same island. 58% of the Galapagos sea lion population is in the southeastern, with the largest rookery in direct contact with the inhabitants. Various strategies have been proposed to reduce the negative impacts of human–animal interaction, ensure the population's viability over time, and reduce the species' extinction risk.

Keywords: Galapagos sea lion, Galapagos fur seal, population trends, anthropogenic disturbances, conservation strategies

1. Introduction

The world's smallest sea lion and fur seal inhabit the Galapagos archipelago, a Natural Heritage of Humanity, located in the Eastern Tropical Pacific at 1000 km off Ecuador's mainland. This ecosystem is exposed to strong environmental variability as a product of local and regional processes that establish unique ecological characteristics that provide the home of a wide diversity of taxa and a high level of endemism [1, 2].

The Galapagos sea lion, *Zalophus wollebaeki*, and Galapagos fur seal, *Arctocephalus galapagoensis*, are endemic otariids of the archipelago that face different challenges to survive [3–5]. Environmental and anthropogenic factors impact the population dynamics of pinnipeds [6], so both species developed adaptation mechanisms allowing them to persist in the archipelago despite the ecosystem's uncertainty in terms of variability in feeding resources and climatic conditions [5, 7] and human-induced pressures that contribute to the deterioration of their habitat [8, 9].

The natural environmental variability of the ecosystem influences the life history strategies, abundance, and distribution of pinnipeds [10–12]. The Galapagos Islands present different levels of productivity and wide seasonal climate variability [13, 14] influenced by oceanographic current systems and upwelling patterns [15, 16] that determine biogeographic patterns within the archipelago [17, 18]. Environmental disturbances such as ENSO (El Niño Southern Oscillation), occurring periodically in the Galapagos, intensify interannual environmental variability [19], causing fluctuations in marine productivity that alter the trophic network and impact the demography of top predators [20, 21]. A situation that is worrying given that the frequency and intensity of these effects increase with global warming [22].

Natural disturbances generate negative effects that are exacerbated when combined with anthropogenic disturbances [23]. In Galapagos, the anthropogenic influence recorded in the last decades is related to local population growth and unsustainable tourism [24]. Among the human-induced pressures are fishing, pollution, marine litter, species introduction, and disease agents, which reduce habitat quality and threaten the health of the species [25–27].

Like other marine predators, the Galapagos pinnipeds serve as indicators of ecosystem health [28, 29]. They depend on terrestrial and marine habitats and are exposed to different stressors [5, 30]. Depending on the rookery location and the level of human interaction, Galapagos pinniped populations will present different growth trends and threat levels, which have important implications for management schemes [8, 21]. We provide an overview of how these species have adapted to the Galapagos environment, facing natural and anthropogenic stressors. We also analyze their population growth trends in the last 21 years and management actions to improve their conservation.

2. Galapagos Sea lions and Fur Seals

Z. wollebaeki and *A. galapagoensis* share some similarities due to their common environment. They inhabit the Galapagos Islands throughout the year [31], and their rookeries are small (<1300 individuals) compared to otariids from high latitudes [21]. They direct their foraging trips to specific areas within the archipelago [11, 32] and have long-lasting lactation periods [7, 33]. They are non-migratory species, although vagrants have been sighted along the coast of South and Central America to Mexico [34, 35].

The Galapagos sea lion is distributed throughout the archipelago, mainly on southeastern islands where 62% of its population is concentrated [9]. It is more numerous than the Galapagos fur seal and has a larger size, approximately 40% larger [7, 36]. It presents a marked sexual size dimorphism. Adult males reach a length of 210 cm and a weight of 200 kg, and females reach a length of 176 cm and a weight of 95 kg [7]. Like other sea lions, they are highly gregarious, polygynous, and territorial [5]. They breed, nurse, rest, and thermoregulate in semiaquatic spaces along beaches on islands and islets [30]. They are philopatric to their reproductive rookeries [31] and feeding areas (**Figure 1**) [28, 37].

Female Galapagos sea lions give birth to one pup after 11 months of gestation between August and January [7], although this varies slightly, being earlier in the western and later in the southeastern region of the archipelago [38]. The maternal care consists of feeding cycles of the female in the sea, which ranges from a few hours to 4 days, and suckling of the pup on land, which lasts on average five hours [38, 39]. Weaning occurs after 18 months of birth and extends to three years [33], so females nurse all their reproductive lives [7].

The Galapagos fur seal breeds on eight main rookeries on western and northern islands, with 95% of its population on Fernandina Island [9], which corresponds to

Figure 1.
Juvenile Galapagos sea lion (Zalophus wollebaeki) swimming in shallow waters in Champion Islet. Photo by Andrés Moreira-Mendieta.

regions with greater productivity and strong upwellings [19]. In seven rookeries, it lives in sympatry with the Galapagos sea lion, where each one has specialized in a specific foraging niche [11, 40, 41]. It presents typical characteristics of fur seals, except for the long investment of mothers in the care of pups [33]. It has a marked sexual dimorphism. Adult males reach a length of 150 cm and a weight of 70 kg, while adult females reach a length of 120 and a weight of 40 kg [36]. They are polygynous and highly philopatric to their reproductive rookeries [31, 42] and their feeding sites [11, 43]. They rest, thermoregulate, and breed on rocky shores near deep productive waters and use crevices, caves, large boulders, and rock ledges for shade and shelter (**Figure 2**) [44].

Between August and November is the Galapagos fur seal reproductive season. Females give birth to one pup yearly after 11 months of gestation [44]. Mothers alternate pup suckling periods (7 days ±1.2 days) and foraging trips (17.9 ± 10.6 h) [32, 38]. Lactation lasts 18–36 months after birth, depending on environmental conditions [44], and mothers often nurse newborns and older young simultaneously [33].

3. Environmental variability

Climate variability strongly influences marine productivity with effects that spread via the food chain to top predators [45, 46]. The oceanic conditions of lower productivity and the high seasonal climate variability recorded in the Galapagos represent an ecological disadvantage for the survival of these pinnipeds compared to others that live at higher latitudes [47].

Some of the adaptations developed by the Galapagos pinnipeds include the reduction of their body size, energy requirements [48], and the extension of the lactation

Figure 2.
Adult male Galapagos fur seal (Arctocephalus galapagoensis) in Cabo Hammond rookery. Photo by Andrés Moreira-Mendieta.

period (2–3 years) [33, 44]. These mechanisms allow them to cope with lower availability of prey in their habitats and increase the chances of survival of the offspring when animals are exposed to food stress [33, 38].

The reduction in prey availability causes animals to increase foraging effort, prolonging the duration of their foraging trips, reaching greater distances, or spending less time in the rookery nursing their young [38, 49]. From January to May (warm season), when food availability is lower, fur seals make more foraging trips and spend less time ashore [38]. Feeding behavior modifications are mainly observed in strong El Niño-years, where female Galapagos fur seals prolong their foraging trips up to three times longer than in a normal year [50].

The El Niño event causes variations in marine productivity, reducing the availability of the main prey of Galapagos pinnipeds, resulting in nutritional stress that is reflected in the mortality rates of their populations, mainly of the offspring [20, 50]. The Galapagos sea lion and fur seal use different foraging niches, and there is no overlap in their diet; however, during anomalous years, both feed on prey of similar trophic levels in the regions where they cohabit [11, 40]. The high flexibility in the feeding behavior of the sea lion allows it to reduce competition with the fur seal, counteract the decrease in food in its habitat, and thus increase its survival [51].

The distinctive environmental characteristics between archipelago regions influence these otariids [52, 53]. Both direct their foraging trips to specific patches within the archipelago depending on the location of their rookeries and the feeding strategy used [32, 37, 41, 51]. The diversification of feeding strategies of the sea lion varies with the rookery and region of the archipelago [54]. It is considered an adaptation to the availability of prey, oceanographic characteristics, and the topography of its feeding sites that allows for reducing competition between individuals from the same rookery [51, 53].

The geographical distribution of the Galapagos sea lion and Galapagos fur seal rookeries, together with adaptation mechanisms, have contributed to the differentiation of genetically structured populations within the archipelago [52, 55–57]. The intra- and inter-specific niche segregation has produced differences between western, central, and southeastern sea lion populations [52, 55]. In contrast, the long-term site fidelity and long-distance foraging trips in fur seals have contributed to differences between populations in the archipelago's west, center, and northeast [56, 57].

The abundance and distribution of Galapagos pinnipeds are influenced by climate variability and intensified by the El Niño event [5]. The intensity, length, and frequency of the El Niño event have increased in the last two decades because of global warming [58–60]. Global warming and the associated climate changes are predictable. In Galapagos marine ecosystems, there is expected a reduction in biodiversity, an increase in sea surface temperature and the thermocline depth, and intensifying upwelling winds [60–62].

These changes will increase species that withstand warming stress, the displacement of endemic species adapted to cold waters by circumtropical species with Panamic and Indo-Pacific affinities, and the tropicalization of the ecosystem by 2050 [60]. Galapagos sea lion and fur seal populations are predicted to decline during strong El Niño events [9, 21], increase the likelihood of disease outbreaks, and increase their vulnerability to the impacts of pollutants [63].

4. Anthropogenic disturbances

As natural disturbances, anthropogenic ones may alter the structure and function of the ecosystems [64]. Oceanic insular ecosystems are susceptible to these effects [23]. Therefore, climate change and other human-induced pressures have degraded marine ecosystems worldwide, creating challenges for species and human societies [65].

The booming tourism industry in the last decades in the Galapagos Islands has led to the rising resident and tourist populations [24]. The human population increase has led to overfishing and the introduction of alien species, greater use of resources and services, and generation of waste and pollution [24, 62]. These problems cause negative impacts on the natural resources of the Galapagos, especially on species exposed to high human interaction [25, 66].

Like other top predators, marine mammals are threatened by varying human-induced pressure levels [29]. In the Galapagos Islands, pinnipeds experience adverse effects related to the alteration of their behavior, deterioration in the quality of their habitat, and dangers to their health and physical integrity [25, 26, 67]. These effects occur due to the interaction of otariids with humans, fisheries, and introduced species, specifically dogs, cats, and rodents [25, 27].

Human presence in sea lion rookeries alters the animals' behavior, both the haul-out behavior and nursing patterns [68–70]. The level of aggression of Galapagos sea lions decreases as the level of exposure to human disturbance increases [26]. This behavior is observed on beaches near San Cristóbal Island town due to the large influx of bathers to which the animals have become accustomed (**Figure 3**). The level of human exposure is unrelated to rookery size [26]; however, may influence the quality of maternal care to pups [68]. The largest rookery of sea lions is, in turn, the most exposed to human presence [9], which, in the long term, could negatively impact the breeding successes of this species because of continuous stress, as has been revealed in other pinnipeds [71, 72].

Figure 3.
Galapagos sea lions from the El Malecón rookery resting on Playa Mann, a highly crowded beach near the town on San Cristóbal Island. Photo by Camila Páez-Riofrío.

The interaction of marine mammals with fisheries poses a threat to these species due to by-catch and vessel strike [73]. Adult Galapagos sea lions are more prone to wounds from propeller strikes when they try to steal fishermen's catch during fishing activities at sea, causing damage to hooks and gillnets [25]. They also cause damage to fishing vessels when they climb onto them to rest during the day, causing them to sink [30]. These adverse effects of fishing by sea lion behavior led fishers to take actions that threaten the species' integrity and protect boats with barbed wire and wood with nails [8].

Marine litter constitutes a significant global threat to biodiversity [74]. Impacts on Galapagos sea lions include cuts, entanglement, strangulation, suffocation, and intestinal obstruction [75]. Plastics are the primary debris type found in oceans affecting otariids [76–79] and are one of the main threats to Galapagos sea lions (**Figure 4**) [79, 80]. This is due to the susceptibility of this species to biomagnification of persistent organic pollutants (POPs), which can be absorbed and transported by microplastics [67, 81].

The Galapagos pinnipeds are exposed to POPs and mercury, which are toxic and bioaccumulative [75, 81]. These compounds impact the immune and endocrine systems of animals, affecting their ability to fight diseases and reproduce successfully [75, 81], especially during periods of nutritional and environmental stress such as those recorded during strong El Niño events [67]. Climate change can exacerbate the effects of pollutants on species, increasing their vulnerability and the risk of reaching their population tipping point [63].

Introduced animals in continuous interaction with endemic species may increase the risk of pathogen spillover in island ecosystems [82, 83]. Diseases with low pathogenicity may become life-threatening for marine mammals during food stress periods [84],

Figure 4.
Galapagos sea lions exposed to plastic residues in Champion Islet, an uninhabited site of Floreana Island. Picture taken during the Galapagos pinniped 2022 annual census. Photo by Marjorie Riofrío-Lazo.

producing significant mortalities in pinnipeds [85]. Galapagos pinnipeds face an increasing threat due to infectious diseases related to domestic animals (dogs, cats, and rats) [27, 83, 86, 87]. Pathogens of canine and feline origin have been reported in the four populated islands of the archipelago [82]. As a consequence, Galapagos sea lions have shown symptoms of infection by Leptospira, Canine distemper virus [88], *Dirofilaria immitis* [27], Mycoplasma [87], and group A rotaviruses [86]. The latter, in turn, is present in Galapagos fur seals [86].

5. Population abundance and conservation status

The Galapagos sea lions and Galapagos fur seals are among the endemic species with conservation priority in the archipelago [9]. They are cataloged in endangered status by the International Union for Conservation of Nature IUCN as their populations have declined over the last decades with no apparent recovery [3, 4, 9].

The causes of population reduction include (1) the environmental variability of the ecosystem [5]; (2) natural disturbances such as the intense El Niño events recorded in 1982/83 and 1997/98 [21, 89] whose residual effects persist; and ((3) anthropogenic disturbances with notable effects in rookeries located on islands with human settlements [8, 21, 89, 90].

Ten censuses were carried out in the entire archipelago from 1978 to 2022 [9, 31, 91]. The correction factors used in the first two censuses (1978 and 2001) for population estimates were not reported. However, from 2014 to the present, this methodology and animal counting have been standardized [5], reducing the uncertainty in population estimates in the last eight years.

The Galapagos sea lion suffered a substantial 50% population decline from 1978 to 2001 (**Table 1**) [3]. Monitoring in the El Malecón rookery in the southeast indicated an annual growth rate of 2% from 2005 to 2015 [21]. El Malecón is the largest rookery in the archipelago, currently making up 16% of the total population (Riofrío-Lazo

Census year	Population counted	Population estimated	Reference
1978	8000	30,000–40,000	Trillmich [31]
2001	4937	14,000–16,000	Alava and Salazar [91]
2014	4980	18,864–26,895	Páez-Rosas, Torres [9]
2015	3791	14,579–20,726	Páez-Rosas, Torres [9]
2016	4309	16,809–23,963	Páez-Rosas, Torres [9]
2017	4344	16,313–22,951	Páez-Rosas, Torres [9]
2018	4891	17,617–24,649	Páez-Rosas, Torres [9]
2019	4435	–	DPNG unpublished data
2021	4561	–	DPNG unpublished data
2022	4385	13,411–18,487	Riofrío-Lazo et al. unpublished data

Table 1.
Galapagos sea lion population abundance in the archipelago per census year.

et al., unpublished data). From 2014 to 2018, the population in the archipelago increased at an average annual rate of 1% [9]. From 2018 to 2022, the number of animals counted decreased by 14% in the archipelago, registering 4385 individuals in 2022 (Riofrío-Lazo et al., unpublished data).

The Galapagos sea lion population in 2022 is estimated at 13,411–18,487 individuals. It has maintained fluctuations in its number over the last eight years, with a slight tendency to decrease from 2014 to 2022 (Riofrío-Lazo et al., unpublished data). The average annual variation rates of the population are different between regions and tend to increase (<6.5%) in the west and north and decrease slightly (<2%) in the center and southeast (Riofrío-Lazo et al., unpublished data, **Figure 5**). The little variability in abundance from 2001 to 2022 and concerning the

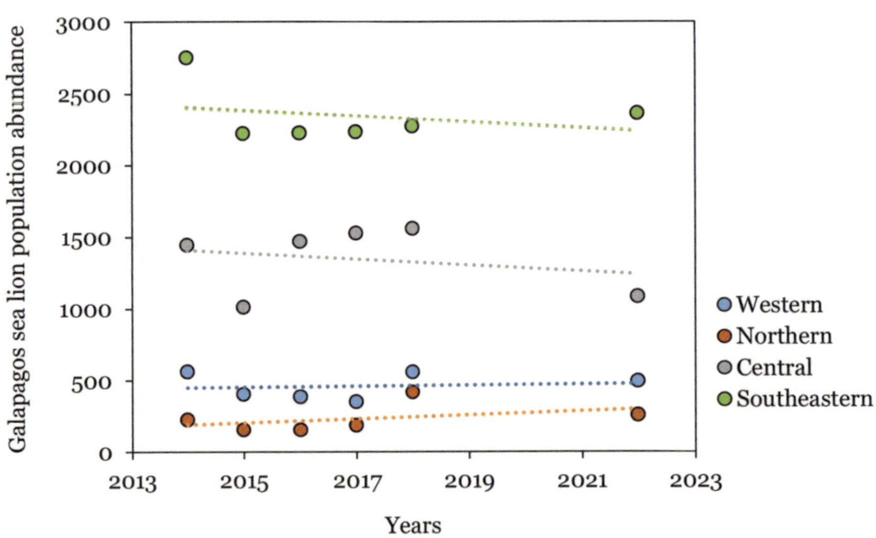

Figure 5.
Population trend (from 2014 to 2022) of the Galapagos sea lion by archipelago region. 2014–2018 census data taken from Páez-Rosas, Torres [9] and 2022 census data (Riofrío-Lazo et al., unpublished data).

Census year	Population counted	Population estimated	Reference
1978	9785	30,000	Trillmich [31]
2001	2733	6000–8000	Alava and Salazar [91]
2002–2005	—	15,000	Salazar and Denkinger [92]
2014	2843	10,676–15,173	Páez-Rosas, Torres [9]
2015	1760	6990–10,230	Páez-Rosas, Torres [9]
2016	2118	7591–10,858	Páez-Rosas, Torres [9]
2017	1996	7110–10,025	Páez-Rosas, Torres [9]
2018	3093	11,615–16,512	Páez-Rosas, Torres [9]
2019	2791	—	DPNG unpublished data
2021	3183	—	DPNG unpublished data
2022	2834	8815–12,005	Riofrío-Lazo et al. unpublished data

Table 2.
Galapagos fur seal population abundance in the archipelago per census year.

first census in 1978 suggests that the reduction of more than 50% of the population persists in the last four generations and that its IUCN conservation status should be kept as endangered.

The Galapagos fur seal suffered a drastic population reduction of 74–80% from 1978 to 2001 (**Table 2**) [4]. Based on extrapolation from 2002 to 2005 primary rookery surveys, it was estimated that the population had decreased by more than 60% from 1978 to 2005 [90]. However, from 2014 to 2018, an average annual increase rate of 3% was estimated [9]. The census 2022 shows values close to those recorded in 2001 [91], with 2834 animals counted.

The increasing trend of the Galapagos fur seal is maintained in the three regions where it lives, with fluctuations over the last 8 years (**Figure 6**). In the west and

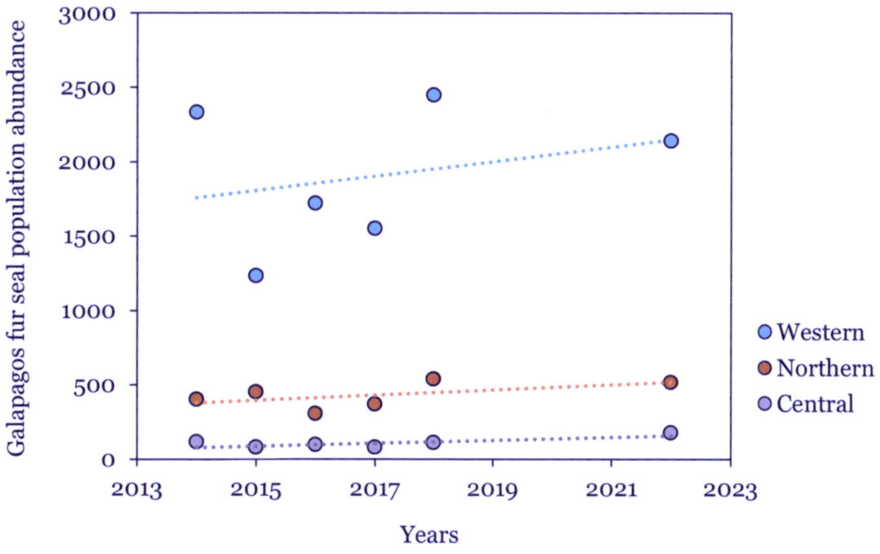

Figure 6.
Population trend (from 2014 to 2022) of the Galapagos fur seal by archipelago region. 2014–2018 census data taken from Páez-Rosas, Torres [9] and 2022 census data (Riofrío-Lazo et al., unpublished data).

north, the population decreased by 12.47% and 3.73% from 2018 to 2022, respectively, while, in the center, the population increased by 58.93% in those years (Riofrío-Lazo et al., unpublished data). Little abundance changes from 2001 to 2022 suggest that the population remains stable, and about the first population monitoring, more than 60% reduction persists in the last four generations (1978–2022). Therefore, its IUCN conservation status should be maintained as endangered.

5.1 Analyzing population trends

Habitat characteristics and population dynamics of Galapagos pinnipeds influence their habitat use in each region [43, 55]. Abundance patterns between regions and between rookeries in the same region of the archipelago are different [9, 21]. Therefore, the population trend of a single rookery should not be considered representative of the trend in the entire region or archipelago [21, 89].

Population trends from 2014 to 2022 of the Galapagos fur seal in the west, center, and north follow the same pattern of increase; however, the average annual growth rates (~3 to 8%) are distinct (Riofrío-Lazo et al., unpublished data). Differences in growth rates show that Galapagos fur seals' most important breeding rookeries are in the west, where greater production of offspring is recorded annually [9], which is related to high-quality habitat and greater availability of prey [11, 41].

The differences in the Galapagos sea lion population trends between regions are explained by various factors: (1) Shelf habitat characteristics that could lead to a restriction of food resources in the north [32]; (2) High levels of marine productivity that promote greater abundance and availability of food but are not used by the sea lion by limiting their feeding effort and the size of their rookery to reduce competition with the fur seal in the west [11, 32]; (3) Different rates of offspring production between rookeries around the archipelago linked to feeding resources available for mothers [28, 51]; and (4) Specific foraging areas with distinct prey species in the diet of individuals for each region [43, 54]. This last explains that sea lions inhabiting less productive areas, such as the center and southeastern, consume prey of high caloric value [43, 51, 93], which allows them to sustain large populations in these regions.

Three of the four populated islands are in the southeast and center; in one of them, San Cristóbal Island, is the El Malecón rookery, which is the most numerous with the highest annual pup production, near the town, and in direct contact with Galapagos residents and tourists [9, 21]. Anthropogenic disturbances to which this population is exposed are domestic animal contact and the potential transmission of infectious diseases [27, 87, 88], wounds resulting from propeller strikes, marine litter, human-fecal contamination by domestic sewage effluents and other sources of pollution [25, 67, 79]. Although the El Malecón rookery has shown a trend of sustained growth in recent years [21], human-induced pressures can have long-term effects on the population, impacting its survival [25, 26]. Due to not all rookeries showing the same growth trend and being exposed to similar problems, for the management actions design, it is relevant to evaluate each rookery independently according to its specific conditions [9].

6. Importance of conservation strategies

Marine protected areas, along with other management tools, are helpful for resource management and biodiversity conservation [2]. However, they should be established based on a comprehensive analysis of the ecosystem functioning [94].

After 22 years of the implementation of the zoning of the Galapagos Marine Reserve, it does not provide sufficient protection for threatened species and key areas for the functioning of the ecosystem, and primary fishing resources are overexploited [62, 94, 95].

Various species are protected for their high economic or esthetic value rather than their importance in nature [96, 97]. From an ecosystem-based management perspective, it is essential to protect species that maintain the ecosystem's functioning, order, and resilience [64]. The Galapagos pinnipeds are charismatic species with high economic and ecological value in the archipelago [64, 98]. This recognition has a great social impact reflected in the attention of its populations regarding management and conservation [8].

The global concern for the protection of Galapagos pinnipeds has led various institutions to carry out outstanding research efforts to understand the population dynamics of these species [9] and establish specific conservation plans for the management of its natural environment [8]. The product of this is the Galapagos pinniped monitoring program developed since 2008 on San Cristóbal Island in the southeastern region, and the global monitoring of these populations through annual research cruises throughout the archipelago (from 2014 to 2022). This knowledge is relevant for decision-makers in establishing effective protected areas that facilitate wildlife management [9].

Each reproductive rookery should be evaluated independently to improve management strategies since population dynamics are variable. The relevance of certain rookeries has been identified by analyzing the population growth trends of both species in different regions [9]. Factors to consider are (1) the high degree of philopatry to its breeding rookeries [99], (2) the high mobility of individuals between rookeries [52, 99] that provide the potential for transmission of infectious diseases, and (3) the level of interaction with humans. Considering these factors, different levels of importance or vulnerability of populations may be determined, and those rookeries with conservation priority to be identified.

Under normal conditions of environmental variability, only two-thirds of all pups survive to the age of one year, and half of the juveniles to the age of two years [100]. An increase in mortality due to human-induced pressures or disease may impede population recovery in the long term [25]. Thus, actions aimed at improving the quality of their habitat, avoiding the deterioration of their health, and raising awareness among the human population about protecting these species must be intensified.

Various strategies have been implemented and suggested to reduce the exposure of sea lions to human presence, avoid the deterioration of the quality of their marine-terrestrial habitats in urban sites as a result of the daily activities of their inhabitants, reduce the negative impact of sea lions on fishing vessels, and reduce their interaction with introduced species; see [9]. However, to achieve these objectives, it is necessary: (1) the joint effort and long-term commitment of the authorities responsible for management, (2) compliance with the regulations by all inhabitants and tourists who visit the Galapagos, (3) the continuation of the annual global censuses of these otariids, and (4) the reinforcement of health studies of their populations, to adopt prevention and early control measures of their infectious agents.

A greater understanding of the threats that the Galapagos pinnipeds face for their survival and monitoring their population allows the establishment of more effective protection measures. The conservation of these species benefits the entire socio-ecosystem of the Galapagos Islands as it promotes a healthier and more resilient ecosystem [64].

Acknowledgements

We acknowledge the financial support from the Universidad San Francisco de Quito and the Galapagos Science Center. We also thank the Galapagos National Park Directorate for the research permit and technical support.

Author details

Marjorie Riofrío-Lazo* and Diego Páez-Rosas
Universidad San Francisco de Quito USFQ and Galapagos Science Center GSC, Galapagos Islands, Ecuador

*Address all correspondence to: mriofriol@usfq.edu.ec

References

[1] Palacios DM, Cantor M. Priorities for ecological research on cetaceans in the Galápagos Islands. Frontiers in Marine Science. 2023;**10**:159

[2] Edgar GJ, Banks S, Bensted-Smith R, Calvopiña M, Chiriboga A, Garske LE, et al. Conservation of threatened species in the Galapagos marine reserve through identification and protection of marine key biodiversity areas. Aquatic Conservation: Marine and Freshwater Ecosystems. 2008;**18**(6):955-968

[3] Trillmich F. *Zalophus wollebaeki*. The IUCN Red List of Threatened Species 2015. 2015.e.T41668A45230540. Available from: https://doi.org/10.2305/IUCN.UK.2015-2.RLTS.T41668A45230540.en [Accessed: October 3, 2023]

[4] Trillmich F. *Arctocephalus galapagoensis*. The IUCN Red List of Threatened Species 2015: . 2015;e.T2057A45223722. Available from: https://doi.org/10.2305/IUCN.UK.2015-2.RLTS.T2057A45223722.en. [Accessed: October 3, 2023]

[5] Riofrío-Lazo M, Páez-Rosas D. Galapagos sea Lions and fur seals, adapted to a variable world. In: Campagna C, Harcourt RG, editors. Ethology and Behavioral Ecology of Otariids and the Odobenid: . Ethology and Behavioral Ecology of Marine Mammals. Cham: Springer; 2021. pp. 643-661

[6] Forcada J, Trathan PN, Reid K, Murphy EJ. The effects of global climate variability in pup production of Antarctic fur seals. Ecology. 2005;**86**:2408-2417

[7] Trillmich F, Jeglinski JW, Meise K, Piedrahita P. The Galapagos sea lion: Adaptation to spatial and temporal diversity of marine resources within the archipelago. In: Denkinger J, Vinueza L, editors. The Galapagos Marine Reserve. Social and Ecological Interactions in the Galapagos Islands. Cham: Springer; 2014. pp. 61-70

[8] Páez-Rosas D, Guevara N. Management strategies and conservation status of Galapagos sea lion populations at San Cristóbal Island, Galapagos, Ecuador. In: Alava JJ, editor. Tropical Pinnipeds: Bio-Ecology, Threats and Conservation. 1st ed. Boca Raton: CRC Press; 2017. pp. 159-175

[9] Páez-Rosas D, Torres J, Espinoza E, Marchetti A, Seim H, Riofrío-Lazo M. Declines and recovery in endangered Galapagos pinnipeds during the El Niño event. Scientific Reports. 2021;**11**(1):8785

[10] Soto KH, Trites AW, Arias-Schreiber M. The effects of prey availability on pup mortality and the timing of birth of south American sea lions (*Otaria flavescens*) in Peru. Journal of Zoology. 2004;**264**(4):419-428

[11] Páez-Rosas D, Aurioles-Gamboa D, Alava JJ, Palacios DM. Stable isotopes indicate differing foraging strategies in two sympatric otariids of the Galapagos Islands. Journal of Experimental Marine Biology and Ecology. 2012;**424**:44-52

[12] Elorriaga-Verplancken FR, Sierra-Rodríguez G, Rosales-Nanduca H, Acevedo-Whitehouse K, Sandoval-Sierra J. Impact of the 2015 El Niño-Southern Oscillation on the abundance and foraging habits of Guadalupe fur seals and California sea lions from the San Benito Archipelago, Mexico. Plos One. 2016;**11**(5):e0155034

[13] Wellington GM, Strong AE, Merlen G. Sea surface temperature

variation in the Galapagos Archipelago: A comparison between AVHRR nighttime satellite data and in situ instrumentation (1982-1998). Bulletin of Marine Science. 2001;**69**(1):27-42

[14] Schaeffer BA, Morrison JM, Kamykowski D, Feldman GC, Xie L, Liu Y, et al. Phytoplankton biomass distribution and identification of productive habitats within the Galapagos Marine Reserve by MODIS, a surface acquisition system, and in-situ measurements. Remote Sensing of Environment. 2008;**112**(6):3044-3054

[15] Sweet W, Morrison JM, Kamykowski D, Schaeffer B, Banks S, McCulloch A. Water mass seasonal variability in the Galápagos Archipelago. Deep Sea Research Part I: Oceanographic Research Papers. 2007;**54**(12):2023-2035

[16] Forryan A, Naveira Garabato AC, Vic C, Nurser AG, Hearn AR. Galápagos upwelling driven by localized wind–front interactions. Scientific Reports. 2021;**11**(1):1277

[17] Edgar G, Banks S, Fariña J, Calvopiña M, Martínez C. Regional biogeography of shallow reef fish and macro-invertebrate communities in the Galapagos archipelago. Journal of Biogeography. 2004;**31**(7):1107-1124

[18] McKinley SJ, Saunders BJ, Rastoin-Laplane E, Salinas-de-León P, Harvey ES. Functional diversity of reef fish assemblages in the Galapagos Archipelago. Journal of Experimental Marine Biology and Ecology. 2022;**549**:151695

[19] Palacios DM, Bograd SJ, Foley DG, Schwing FB. Oceanographic characteristics of biological hot spots in the North Pacific: A remote sensing perspective. Deep Sea Research Part II: Topical Studies in Oceanography. 2006;**53**(3-4):250-269

[20] Trillmich F, Dellinger T. The effects of El Nino on Galapagos pinnipeds. In: Trillmich F, Ono K, editors. Pinnipeds and El Niño: Responses to Environmental Stress. New York: Springer-Verlag; 1991. pp. 66-74

[21] Riofrío-Lazo M, Arreguín-Sánchez F, Páez-Rosas D. Population abundance of the endangered Galapagos sea lion *Zalophus wollebaeki* in the southeastern Galapagos archipelago. PLoS One. 2017;**12**(1):e0168829

[22] Dueñas A, Jiménez-Uzcátegui G, Bosker T. The effects of climate change on wildlife biodiversity of the Galapagos Islands. Climate Change Ecology. 2021;**2**:100026

[23] Russell JC, Kueffer C. Island biodiversity in the Anthropocene. Annual Review of Environment and Resources. 2019;**44**:31-60

[24] Mena CF, Quiroga D, Walsh SJ. Threats to sustainability in the Galapagos Islands: A social–ecological perspective. In: Sarmiento FO, Florich LM, editors. The Elgar Companion to Geography, Transdisciplinarity and Sustainability. Cheltenham: Edward Elgar Publishing Limited; 2020. pp. 342-358

[25] Denkinger J, Gordillo L, Montero-Serra I, Murillo JC, Guevara N, Hirschfeld M, et al. Urban life of Galapagos sea lions (*Zalophus wollebaeki*) on San Cristobal Island, Ecuador: Colony trends and threats. Journal of Sea Research. 2015;**105**:10-14

[26] Walsh JT, Kovaka K, Vaca E, Weisberg DS, Weisberg M. The effects of human exposure on Galápagos sea lion behavior. Wildlife Biology. 2020;**2020**(4):1-8

[27] Gregory TM, Livingston I, Hawkins EC, Loyola A, Cave A, Vaden SL, et al. Dirofilaria immitis identified in Galapagos sea lions (*Zalophus wollebaeki*): A wildlife health and conservation concern. Journal of Wildlife Diseases. 2023;**59**(3):487-494

[28] Drago M, Franco-Trecu V, Cardona L, Inchausti P, Tapia W, Páez-Rosas D. Stable isotopes reveal long-term fidelity to foraging grounds in the Galapagos sea lion (*Zalophus wollebaeki*). PLoS One. 2016;**11**(1):e0147857

[29] Alves F, Rosso M, Li S, Nowacek DP. A sea of possibilities for marine megafauna. Science. 2022;**375**(6579):391-392

[30] Montero-Serra I, Páez-Rosas D, Murillo JC, Vegas-Vilarrúbia T, Fietz K, Denkinger J. Environment-driven changes in terrestrial habitat use and distribution of the Galapagos sea lion. Endangered Species Research. 2014;**24**(1):9-19

[31] Trillmich F. Galapagos sea lions and fur seals. Noticias de Galápagos. 1979;**29**:8-14

[32] Villegas-Amtmann S, Jeglinski JW, Costa DP, Robinson PW, Trillmich F. Individual foraging strategies reveal niche overlap between endangered Galapagos pinnipeds. PLoS One. 2013;**8**(8):e70748

[33] Trillmich F, Wolf JB. Parent–offspring and sibling conflict in Galápagos fur seals and sea lions. Behavioral Ecology and Sociobiology. 2008;**62**:363-375

[34] Páez-Rosas D, Valdovinos LA, Elorriaga-Verplancken FR. Northernmost record of the Galapagos fur seal (*Arctocephalus galapagoensis*): A consequence of anomalous warm conditions around the Galapagos Archipelago. Aquatic Mammals. 2017;**43**(6):629-634

[35] Elorriaga-Verplancken FR, Morales-Vázquez JE, Ortega-Ortiz CD, Llamas-González M, Meza-Yáñez R, Páez-Rosas D. Northernmost Record of the Galapagos sea lion (*Zalophus wollebaeki*): Sightings along the Mexican Central Pacific and the Gulf of California during La Niña conditions. Aquatic Mammals. 2022;**48**(6):478-484

[36] Trillmich F. Maternal investment and sex-allocation in the Galapagos fur seal, *Arctocephalus galapagoensis*. Behavioral Ecology and Sociobiology. 1986;**19**:157-164

[37] Villegas-Amtmann S, Costa DP, Tremblay Y, Salazar S, Aurioles-Gamboa D. Multiple foraging strategies in a marine apex predator, the Galapagos sea lion *Zalophus wollebaeki*. Marine Ecology Progress Series. 2008;**363**:299-309

[38] Trillmich F. Attendance behavior of Galapagos fur seals. In: Gentry RL, Kooyman GL, editors. Fur Seals: Maternal Strategies on Land and at Sea. Princeton, NJ: Princeton University Press; 1986. pp. 168-185

[39] Villegas-Amtmann S, McDonald BI, Páez-Rosas D, Aurioles-Gamboa D, Costa DP. Adapted to change: Low energy requirements in a low and unpredictable productivity environment, the case of the Galapagos sea lion. Deep Sea Research Part II: Topical Studies in Oceanography. 2017;**140**:94-104

[40] Dellinger T, Trillmich F. Fish prey of the sympatric Galapagos fur seals and sea lions: Seasonal variation and niche separation. Canadian Journal of Zoology. 1999;**77**:1204-1216

[41] Jeglinski JW, Goetz KT, Werner C, Costa DP, Trillmich F. Same size–same niche? Foraging niche separation between sympatric juvenile Galapagos

sea lions and adult Galapagos fur seals. Journal of Animal Ecology. 2013;**82**:694-706

[42] Bonner WN. Seals of the Galapagos Islands. Biological Journal of the Linnean Society. 2008;**21**(1-2):177-184

[43] Páez-Rosas D, Rodríguez-Pérez M, Riofrío-Lazo M. Competition influence in the segregation of the trophic niche of otariids: A case study using isotopic Bayesian mixing models in Galapagos pinnipeds. Rapid Communications in Mass Spectrometry. 2014b;**28**:2550-2558

[44] Trillmich F. Galapagos fur seal, *Arctocephalus galapagoensis*. In: Croxall JP, Gentry RL, editors. Status, Biology, and Ecology of fur Seals. NOAA Technical Report NMFS 51. Seattle, Washington: US Department of Commerce; 1987. pp. 23-28

[45] Doney SC, Ruckelshaus M, Emmett Duffy J, Barry JP, Chan F, English CA, et al. Climate change impacts on marine ecosystems. Annual Review of Marine Science. 2012;**4**:11-37

[46] Wernberg T, Smale DA, Tuya F, Thomsen MS, Langlois TJ, De Bettignies T, et al. An extreme climatic event alters marine ecosystem structure in a global biodiversity hotspot. Nature Climate Change. 2013;**3**(1):78-82

[47] Chilvers BL, Robertson BC, Wilkinson IS, Duignan PJ. Growth and survival of New Zealand sea lions, *Phocarctos hookeri*: Birth to 3 months. Polar Biology. 2007;**30**:459-469

[48] Trillmich F, Kooyman GL. Field metabolic rate of lactating female Galápagos fur seals (*Arctocephalus galapagoensis*): The influence of offspring age and environment. Comparative Biochemistry and Physiology Part A: Molecular & Integrative Physiology. 2001;**129**(4):741-749

[49] Costa DP. Energetics. In: Perrin WF, Wursig B, Thewissen JGM, editors. Encyclopedia of Marine Mammals. San Diego: Academic; 2009. pp. 383-391

[50] Trillmich F, Limberger D. Drastic effects of El Niño on Galapagos pinnipeds. Oecologia. 1985;**67**:19-22

[51] Páez-Rosas D, Villegas-Amtmann S, Costa D. Intraspecific variation in feeding strategies of Galapagos sea lions: A case of trophic specialization. PLoS One. 2017;**12**(10):e0185165

[52] Wolf JB, Harrod C, Brunner S, Salazar S, Trillmich F, Tautz D. Tracing early stages of species differentiation: Ecological, morphological and genetic divergence of Galápagos sea lion populations. BMC Evolutionary Biology. 2008;**8**:1-14

[53] Páez-Rosas D, Riofrío-Lazo M, Aurioles-Gamboa D. Flexibility in the foraging strategies of the Galapagos sea lion inferred from a multiple approach analysis. In: Denkinger J, Vinueza L, editors. The Galapagos Marine Reserve, Social and Ecological Interactions in the Galapagos Islands. Cham: Springer; 2014. pp. 71-80

[54] Páez-Rosas D, Aurioles-Gamboa D. Spatial variation in the foraging behaviour of the Galapagos sea lions (*Zalophus wollebaeki*) assessed using scat collections and stable isotope analysis. Journal of the Marine Biological Association of the United Kingdom. 2014;**94**(06):1099-1107

[55] Jeglinski JW, Wolf JB, Werner C, Costa DP, Trillmich F. Differences in foraging ecology align with genetically divergent ecotypes of a highly mobile

marine top predator. Oecologia. 2015;**179**:1041-1052

[56] Chaves JA, Lopes F, Martínez D, Cueva DF, Gavilanes GI, Bonatto SL, et al. Population genetics and phylogeography of Galapagos fur seals. Frontiers in Genetics. 2022;**13**:725772

[57] Lopes F, Hoffman JI, Valiati VH, Bonatto SL, Wolf JBW, Trillmich F, et al. Fine-scale matrilineal population structure in the Galapagos fur seal and its implications for conservation management. Conservation Genetics. 2015;**16**(5):1099-1113

[58] Thompson DM, Conroy JL, Collins A, Hlohowskyj SR, Overpeck JT, Riedinger-Whitmore M, et al. Tropical Pacific climate variability over the last 6000 years as recorded in Bainbridge Crater Lake, Galápagos. Paleoceanography. 2017;**32**(8):903-922

[59] Rustic GT, Koutavas A, Marchitto TM, Linsley BK. Dynamical excitation of the tropical Pacific Ocean and ENSO variability by little ice age cooling. Science. 2015;**350**(6267):1537-1541

[60] Banks S, Edgar G, Glynn P, Kuhn A, Moreno J, Ruiz D, et al. A review of Galápagos marine habitats and ecological processes under climate change scenarios. In: Larrea Oña I, Di Carlo G, editors. Climate Change Vulnerability Assessment of the Galápagos Islands. USA: WWF and Conservation International; 2011. pp. 47-68

[61] Larrea Oña I, Di Carlo G, editors. Climate Change Vulnerability Assessment of the Galapagos Islands. USA: WWF and Conservation International; 2011. 116p

[62] Escobar-Camacho D, Rosero P, Castrejón M, Mena CF, Cuesta F. Oceanic islands and climate: Using a multi-criteria model of drivers of change to select key conservation areas in Galapagos. Regional Environmental Change. 2021;**21**(2):47

[63] Alava JJ, McMullen K, Jones J, Barragán-Paladines MJ, Hobbs C, Tirape A, et al. Multiple anthropogenic stressors in the Galápagos Islands' complex social–ecological system: Interactions of marine pollution, fishing pressure, and climate change with management recommendations. Integrated Environmental Assessment and Management. 2023;**19**(4):870-895

[64] Riofrío-Lazo M, Reck G, Páez-Rosas D, Zetina-Rejón MJ, Del Monte-Luna P, Reyes H, et al. Food web modeling of the southeastern Galapagos shelf ecosystem. Ecological Indicators. 2021;**132**:108270

[65] Hoegh-Guldberg O, Bruno JF. The impact of climate change on the world's marine ecosystems. Science. 2010;**328**(5985):1523-1528

[66] Nieto-Claudin A, Deem SL, Rodríguez C, Cano S, Moity N, Cabrera F, et al. Antimicrobial resistance in Galapagos tortoises as an indicator of the growing human footprint. Environmental Pollution. 2021;**284**:117453

[67] Alava JJ, Ross PS. Chapter 8 – pollutants in tropical marine mammals of the Galápagos Islands, Ecuador: An Ecotoxicological quest to the Last Eden. In: Fossi MC, Panti C, editors. Marine Mammal Ecotoxicology. London: Academic Press; 2018. pp. 213-234

[68] Kovacs K, Innes S. The impact of tourism on harp seals (*Phoca groenlandica*) in the Gulf of St. Lawrence, Canada. Applied Animal Behaviour Science. 1990;**26**(1-2):15-26

[69] Suryan R, Harvey J. Variability in reactions of Pacific harbor seals, *Phoca vitulina* richardsi, to disturbance. Fishery Bulletin. 1999;**97**(2):332-339

[70] Kucey L, Trites AW. A review of the potential effects of disturbance on sea lions: Assessing response and recovery. In: Trites AW, Atkinson SK, DeMaster DP, Fritz LW, Gelatt TS, Rea LD, Wynne KM, editors. Sea Lions of the World. Fairbanks: Alaska Sea Grant College Program; 2006. pp. 581-589

[71] Johnson WM, Lavigne DM. Mass tourism and the Mediterranean monk seal. The Monachus Guardian. 1999;**2**(2):1-30

[72] French SS, Gonzalez-Suarez M, Young JK, Durham S, Gerber LR. Human disturbance influences reproductive success and growth rate in California sea lions (*Zalophus californianus*). PLoS One. 2011;**6**(3):e17686

[73] Schipper J, Chanson JS, Chiozza F, Cox NA, Hoffmann M, Katariya V, et al. The status of the world's land and marine mammals: Diversity, threat, and knowledge. Science. 2008;**322**(5899):225-230

[74] Alfaro-Núñez A, Astorga D, Cáceres-Farías L, Bastidas L, Soto Villegas C, Macay K, et al. Microplastic pollution in seawater and marine organisms across the tropical Eastern Pacific and Galápagos. Scientific Reports. 2021;**11**(1):6424

[75] Alava JJ, Palomera C, Bendell L, Ross PS. Pollution as an emerging threat for the conservation of the Galapagos Marine Reserve: Environmental impacts and management perspectives. In: Denkinger J, Vinueza L, editors. The Galapagos Marine Reserve. Social and Ecological Interactions in the Galapagos Islands. Cham: Springer; 2014. pp. 247-283

[76] Perez-Venegas DJ, Toro-Valdivieso C, Ayala F, Brito B, Iturra L, Arriagada M, et al. Monitoring the occurrence of microplastic ingestion in Otariids along the Peruvian and Chilean coasts. Marine Pollution Bulletin. 2020;**153**:110966

[77] Ortega-Borchardt JÁ, Ramírez-Álvarez N, Mendoza LMR, Gallo-Reynoso JP, Barba-Acuña ID, García-Hernández J, et al. Detection of microplastic particles in scats from different colonies of California sea lions (*Zalophus californianus*) in the Gulf of California. Mexico: A preliminary study. Marine Pollution Bulletin. 2023;**186**:114433

[78] Donohue MJ, Masura J, Gelatt T, Ream R, Baker JD, Faulhaber K, et al. Evaluating exposure of northern fur seals, *Callorhinus ursinus*, to microplastic pollution through fecal analysis. Marine Pollution Bulletin. 2019;**138**:213-221

[79] Moreira-Mendieta A, Garcia-Garin O, Muñoz-Pérez JP, Urquía DO, Drago M, Borrell A, et al. Detection and quantification of microplastic pollution in the endangered Galapagos sea lion. Science of the Total Environment. 2023;**896**:166223

[80] Muñoz-Pérez JP, Lewbart GA, Alarcón-Ruales D, Skehel A, Cobos E, Rivera R, et al. Galápagos and the plastic problem. Frontiers in Sustainability. 2023;**4**:18

[81] Alava JJ, Salazar S, Cruz M, Jiménez-Uzcátegui G, Villegas-Amtmann S, Paéz-Rosas D, et al. DDT strikes back: Galapagos sea lions face increasing health risks. Ambio. 2011;**40**:425-430

[82] Culda CA, Dionnet R, Barbu AC, Cârstolovean AS, Dan T, Grijalva J, et al.

The presence of *Dirofilaria immitis* in Domestic Dogs on San Cristobal Island, Galapagos. Pathogens. 2022;**11**(11):1287

[83] Brock P, Hall A, Goodman S, Cruz M, Acevedo-Whitehouse K. Applying the tools of ecological immunology to conservation: A test case in the Galapagos sea lion. Animal Conservation. 2013;**16**(1):19-31

[84] Simeone CA, Gulland FM, Norris T, Rowles TK. A systematic review of changes in marine mammal health in North America, 1972-2012: The need for a novel integrated approach. PLoS One. 2015;**10**(11):e0142105

[85] Osterhaus A, Groen J, Spijkers H, Broeders H, UytdeHaag F, de Vries P, et al. Mass mortality in seals caused by a newly discovered virus-like morbillivirus. Veterinary Microbiology. 1990;**23**(1-4):343-350

[86] Coria-Galindo E, Rangel-Huerta E, Verdugo-Rodríguez A, Brousset D, Salazar S, Padilla-Noriega L. Rotavirus infections in Galapagos sea lions. Journal of Wildlife Diseases. 2009;**45**(3):722-728

[87] Sarzosa MS, Duignan P, DeRango EJ, Field C, Ríos C, Sanchez S, et al. Occurrence of mycoplasmas in Galapagos sea Lions (*Zalophus wollebaeki*) and their association with other respiratory pathogens. The Journal of Wildlife Diseases. 2021;**57**(3):623-627

[88] Denkinger J, Guevara N, Ayala S, Murillo JC, Hirschfeld M, Montero-Serra I, et al. Pup mortality and evidence for pathogen exposure in Galapagos sea lions (*Zalophus wollebaeki*) on San Cristobal Island, Galapagos, Ecuador. Journal of Wildlife Diseases. 2017;**53**(3):491-498

[89] Trillmich F, Meise K, Kalberer S, Mueller B, Piedrahita P, Pörschmann U, et al. On the challenge of interpreting census data: Insights from a study of an endangered pinniped. PLoS One. 2016;**11**(5):e0154588

[90] Alava JJ, Cheung WWL, Ross PS, Sumaila UR. Climate change–contaminant interactions in marine food webs: Toward a conceptual framework. Global Change Biology. 2017;**23**(10):3984-4001

[91] Alava JJ, Salazar S. Status and conservation of Otariids in Ecuador and the Galapagos Islands. In: Trites AW, Atkinson SK, DeMaster DP, Fritz LW, Gelatt TS, Rea LD, Wynne KM, editors. Sea Lions of the World. Fairbanks: Alaska Sea Grant College Program; 2006. pp. 495-519

[92] Salazar S, Denkinger J. Possible effects of climate change on the populations of Galapagos pinnipeds. Galapagos Research. 2010;**67**:45-49

[93] Blakeway J-A, Arnould JP, Hoskins AJ, Martin-Cabrera P, Sutton GJ, Huckstadt LA, et al. Influence of hunting strategy on foraging efficiency in Galapagos sea lions. PeerJ. 2021;**9**:e11206

[94] Riofrío-Lazo M, Zetina-Rejón MJ, Reck G, Páez-Rosas D, Arreguín-Sánchez F. Challenges in the application of the ecosystem approach to fisheries Management in the Galapagos Islands. In: Walsh SJ, Mena CF, Stewart JR, Muñoz Pérez JP, editors. Island Ecosystems: Challenges to Sustainability. Cham: Springer International Publishing; 2023. pp. 319-334

[95] Moity N. Evaluation of No-take zones in the Galapagos Marine Reserve, Zoning Plan 2000. Frontiers in Marine Science. 2018;**5**:244

[96] Jordán F. Keystone species and food webs. Philosophical Transactions of the

Royal Society B: Biological Sciences. 2009;**364**(1524):1733-1741

[97] Simberloff D. Flagships, umbrellas, and keystones: Is single-species management passé in the landscape era? Biological Conservation. 1998;**83**(3):247-257

[98] Kovacs KM, Aguilar A, Aurioles D, Burkanov V, Campagna C, Gales N, et al. Global threats to pinnipeds. Marine Mammal Science. 2012;**28**(2):414-436

[99] Meise K, Krüger O, Piedrahita P, Trillmich F. Site fidelity of male Galápagos sea lions: A lifetime perspective. Behavioral Ecology and Sociobiology. 2013;**2013**:1-11

[100] Kraus C, Mueller B, Meise K, Piedrahita P, Pörschmann U, Trillmich F. Mama's boy: Sex differences in juvenile survival in a highly dimorphic large mammal, the Galapagos sea lion. Oecologia. 2012;**171**:893-903

Chapter 2

The Critically Endangered Pampa Cat (*Leopardus munoai*) on the Brink of Extinction in Brazil: The Little We Know and an Action Plan to Try to Save It

Fábio Dias Mazim, Paulo Guilherme Carniel Wagner, Lester Alexander Fox-Rosales, Alisson da Rosa Boÿink and Tadeu Gomes de Oliveira

Abstract

Leopardus munoai, the Pampa cat, is a small felid (ca. 3–4 kg) recently separated from the *Leopardus colocola* complex. This Pampa cat is endemic to the Pampas grasslands of Uruguay, western Argentina and southernmost Brazil. Originally, the Pampas had a long history of extensive cattle ranching; however, its fields have been exponentially converted mostly to soy fields in the last five decades. Species distribution models have shown a reduction in their area (1997–2022) of 42%, with only 31,808 km^2 of suitable habitat remaining. Despite the Brazilian Pampas being surveyed in their entirety, no records of this felid have ever been repeated at the same site in different years since 1997 (50,000 trap-nights; >400,000 km of highway day/night crossing). This effort generated only 32 records (17 road-kills). Despite the huge detection efforts, there has never been a confirmed resident population detected anywhere. The scenario that unfolds is of an extremely rare felid that seems to use the vanishing native herbaceous fields, which are dependent upon cattle grazing, in replacement of the original (now extinct) grazing megafauna. With an expected population of 45–50 individuals, which has become virtually extinct in its original eastern portion, all signs point towards imminent extinction.

Keywords: *Leopardus munoai*, extinction, natural history, population assessment, action plan, Pampa

1. Introduction

The Neotropical region presents a rich diversity of felines with many species living in sympatry. In the temperate fields of the Brazilian Pampas, the usual feline assemblage is composed of the Geoffroy's cat (*Leopardus geoffroyi* d'Orbigny & Gervais, 1844), the Pampas cat (*Leopardus colocola* Molina, 1782), the margay (*Leopardus wiedii* Schinz, 1821) and the jaguarundi (*Herpailurus yagouaroundi* Saint-Hilaire, 1803) [1]. This set of species is unique in the world due to the Pampas being the limits of the northeastern distribution of Geoffroy's cat, the southern distribution of margay and jaguarundi and a place of occurrence of a disjunct population of Pampas cat that is recognized as a species/subspecies *L.* (*colocola*) *munoai* [2, 3]. Here, we use the term Pampas cat in Ref. to the original broader group, i.e., *L. colocola* group/complex, and Pampa cat to refer specifically to the species that is endemic to the Pampas biome, i.e., *Leopardus munoai*.

The Pampa cat (*Leopardus munoai*) – **Figure 1**, also known as the Uruguayan pampas cat, is a small wild felid that almost exclusively inhabits the fields of the Uruguayan savanna, a subtropical native grassland ecoregion [4] that is located in Uruguay, the extreme south of Brazil (state of Rio Grande do Sul – restricted area to the Pampas biome), as well as a small area in northeastern Argentina. This field seems to have been historically isolated in this region, possibly due to geographical barriers such as the Plata River to the south, the Paraná/Paraguay rivers to the

Figure 1.
Leopardus munoai, the Pampa cat. Top and right-side pictures by Caio Sarmento Belleza.

west and the Atlantic Forest to the north [3, 5–7]. For Martínez-Lanfranco and González [8], the term pampas cat has traditionally been used to refer to small, non-spotted, Neotropical felines of the genus *Leopardus* Gray, 1842, which, unlike their spotted congeners that mainly use forest habitats, occur in open environments throughout South America [3, 9].

The region where the Pampa cat lives are formed by a complex of fields that are native to the southern region of the state of Rio Grande does Sul (extreme south of Brazil on the border with Argentina and Uruguay), which is both officially and popularly named as the Pampas biome and is characterized by elements of the Austral-Antarctic flora [10–12]. In addition, due to the fertile and deep soil favorable for agriculture, this ecoregion is considered one of the most critical conservation priorities for terrestrial vertebrates in the Neotropics [13]. Given this scenario, the Pampa cat has been classified as 'endangered' (EN) of extinction in the regional listing of the state of Rio Grande do Sul [14] and as a threatened species with conservation priority in Uruguay [15]. These categorizations, as well as a general overall assessment for the complex *L. colocola* [16], were performed before this endemic felid was formally recognized as a distinct taxon [3] and were based on ecological data that was more scarcer than what is currently available. In Brazil, it was recently classified as *L. munoai* and placed in the critically endangered (CR) category. As an aggravating factor, when reviewing the conservation status of this small felid as well as the conservation status of the species' native fields, Tirelli et al. [17] showed that the 24 demographic scenarios evaluated generated an estimated population size that places the Pampa cat in one of the IUCN in danger of extinction categories. Furthermore, eight scenarios placed it in the endangered category (EN) and five scenarios in the critically endangered category (CR), which indicates that the species may be seriously endangered.

2. Brief taxonomic overview

Despite previous taxonomic assessments [9] there are currently two taxonomic proposals within the *Leopardus colocola* complex. Kitchener's who evaluated previous studies and took into account biogeographic, morphological and genetic aspects, recognized only one species with seven subspecies [2]: *Leopardus colocola colocola, Le. c. wolffsohni, Le. c. pajeros* (including *crucina*), *Le. c. budini* (including *steinbachi*), *Le. c. garleppi* (including *Thomasi*), *Le. c. bracatus*, and *Le. c. munoai*. Nascimento et al. [3] using the widest morphological coverage of the *Le. colocola* complex at that moment, combined with multilocus phylogeny, species delimitation techniques, and ecological niche analyses, stated that treating pampas cats as a single species would underestimate their actual diversity. When analyzing the distribution and frequency of characters along the entire distribution of the pampas cat to detect sharp clines and discontinuities and take into account intra- and interpopulational variation, which are key factors for defining taxa and searching for diagnostic characters [18–20], the authors claim to recognize five allopatric groups, in which each would have clear diagnostic characteristics and well-defined geographical distribution. The oldest epithets available for the groups would be: Group I = *colocola* Molina, 1782; Group II = *garleppi* Matschie, 1912; Group III = *pajeros* Desmarest, 1816; Group IV = *braccata* Cope, 1889; and Group V = *muñoai* Ximénez, 1961. Overall, these analyses demonstrated that specimens with this genotype/phenotype have unique morphological, genetic, and ecological characteristics, and they should be considered a distinct unit for conservation assessment and management actions. *Leopardus munoai* is a species that occurs in rural areas with heavy rainfall in the dry season in southern Brazil, Uruguay and northeastern Argentina. There is an urgent need to carry out such conservation planning on behalf of this endemic felid, since the native fields of the Uruguayan savanna have been largely transformed by agriculture (especially rice and soybean crops) and forest plantations (*Eucalyptus* spp. and *Pinus* spp.) [21, 22]. These human activities have led to the extinction of several species of local mammals [23], although not endemic species such as the Pampa cat.

Note it has been argued recently that the species name should change from *Leopardus munoai* [24], to *Leopardus fasciatus* [25], which is likely to happen [8]. We are not discussing the merit; we are simply using the most currently recognizable and used name in this chapter, *Leopardus munoai*.

3. Aims and scope

Given the current understanding of its high degree of endemism, taxonomic distinction, scarcity of refined ecological data, and successive exponential reduction of its habitat, which demonstrate an increasingly real scenario of the threat of extinction, we consider it appropriate to present an in-depth assessment of its little-known natural history and conservation situation. To this end, we seek to refine the occurrence records, relate them to the analysis of suitability and habitat loss, and make a profound analysis of the causal factors that have led the species to the threshold of extinction in Brazil, in order to present a proposal for an emergency action plan to try to save it.

4. The setting

The Pampas biome, which has approximately 178,000 km^2, corresponds to 2.07% of the Brazilian territory, and is restricted to the state of Rio Grande do Sul, though it is the dominant landscape in 63% of the territory of the state [26]. This domain is covered by four geomorphological regions, specifically the Gaúcha Central Depression, the Rio Grande do Sul Shield (Serra do Sudeste), the Coastal Plain and the Plateau of the Campanha [26] – **Figure 2**.

The climate, according to the classification of Köppen [27], is of the humid subtropical type, with precipitation equally distributed throughout the year (Cfa and Cfb). The average temperature for the entire study area varies between 17 and 18°C, where maximum temperatures can exceed 40°C and minimum temperatures fall below −5°C.

The pastures of the region are vegetable formations that are conditioned, mainly, by edaphic characteristics and not so much climatic ones [28, 29]. In addition to edaphic factors, grazing pressure and conversion to agricultural or forestry areas strongly influence these rural formations, thus creating different physiognomies (fields with low vegetation; fields with tall vegetation, subshrubbed or scrubby fields; and mixed fields, fields of *Baccharis dracunculifolia* or paleaceous fields), as described by several authors [28–30]. Recently, based on climatic conditions, floristic composition and edaphic characteristics, Menezes [12] concluded that the fields of the Pampas biome could be subdivided into three distinct groups: Pampas Mesic fields, Coastal Pampas and Coastal Pampas pasture.

According to Marchiori [31], the native vegetation of the southwest, which consists of extensive rural areas with forests on the slopes of sandstone plateaus and

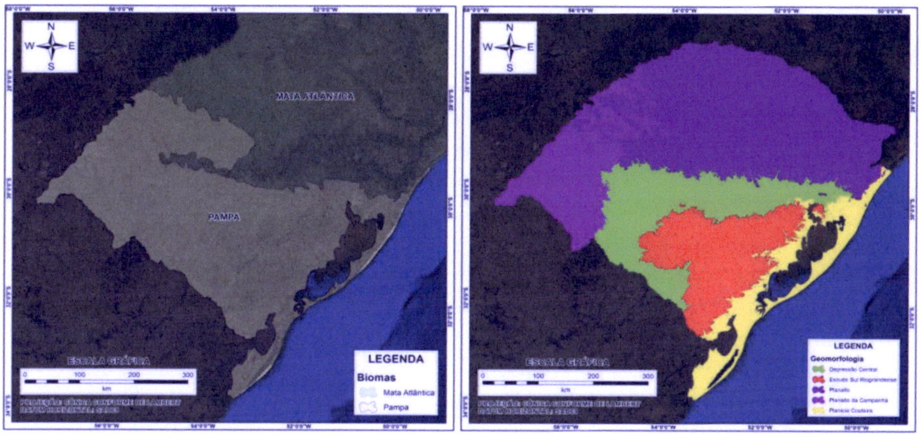

Figure 2.
Map showing the limits of the pampas biome and the geomorphologic regions of the state of Rio Grande do Sul, Brazil.

river banks and park-like formation, still presents areas in its original state, although the complications promoted by anthropic actions have spread exponentially. Verdum [32] reports the natural process of the sandification of the fields, which is currently accelerated, and that the fields are subject to degradation by inappropriate land use such as agrarian expansion and implementation of monocultures.

4.1 Data collection

All possible types of records were compiled from those used in surveys of Neotropical carnivorous mammals. These included the use of camera traps, capture traps for radio collar placement and collection of biological material, recording footprints and faeces, random visual observations, and/or in night-time transections. A search was also carried out for individuals run over on highways in the Pampa biome or killed specimens (firearms or domestic dogs) as a form of retaliation for attacking poultry farms [33].

The information collected, provided by third parties (colleagues and farmers) and systematic searches for the Pampa cat began in 1997. This generated an effort of 50,000 trap-nights, 7000 hours of active searches in night transections, 1100 live-catch traps/night, in addition to 300,000 kilometers of highways traveled in the search for individuals that had been run over. Camera trapping followed standard and well-established protocols that targeted small cats (for a detailed description see [34–36]). In 2021, at the Saicã site, which covers 51,000 hectares and is located between the municipalities of Rosario do Sul and Cacequi (RS), we deployed 6–9 camera traps spaced at ca. 500 m (400–600 m) from each other that focused on the native fields that have been free from agriculture for at least 40 years. Thus, cameras were placed specifically targeting the Pampa cat's presumed preferred habitat.

Live-captures followed strict protocols, such as those of the American Society of Mammalogists [37], all with proper licenses from environmental agencies (IBAMA/SEMA), none of them were conducted for the purpose of this work, and all of them to attend the legal requirements of the aforementioned environmental agencies. All records refer exclusively to the Brazilian Pampas biome (part of the Uruguayan savanna ecoregion in Brazil). All the municipalities that belong to the domains of the

Brazilian Pampas were included in the searches, with the area covered and time effort varying between areas. Additionally, in 2023, all of the points where the records of the species occurred between 1997 and 2022 were visited again, as a form of in-person assessment of the state of conservation of the habitats. In order to determine the relative abundance index (RAI), we considered only one record/hour [38].

4.2 Data analysis

We modeled the habitat suitability of the Pampa cat in Rio Grande do Sul state using the maximum entropy algorithm in Maxent 3.4.4 [39]. This algorithm uses presence-only data as well as background points obtained from a geographical area defined *a priori*. In our case, we defined our sampling area as the areas of Rio Grande do Sul state in the Uruguayan savanna ecoregion [4]. Using the species locations and the covariate values, the algorithm estimates the target probability distribution by using the principle of maximum entropy (i.e., finding the distribution that is closest to uniform). Maxent has been widely applied for modeling species distribution and habitat suitability, and thereby identifies priority areas for conservation as well as delineating species ranges.

We contemplated 19 bioclimatic variables [40] and land cover [41]. All bioclimatic predictors were checked for multicollinearity, and a PCA was performed to select the three variables that maximized variability among the locations where the Pampa cat was registered. The variables selected were mean annual precipitation (bio12), annual temperature range (bio7), and isothermality (bio3). Since our records date back to 1997, and given the fast rate of habitat loss in the Brazilian Pampas, suitability for the species was modeled using land cover data for the year 1997. We then estimated the current suitable habitat that is left by removing areas that were transformed between 1997 and 2021.

5. Natural history traits of an almost extinct species

5.1 Diet

Analysis of the stomach contents of 14 specimens that had been run over [42] indicated that rodents represent 63.2% of the diet, followed by birds (31.6%) and amphibians (5.2%). Of the species of prey, Brazilian guinea pigs (*Cavia aperea* 549 g) comprised 33.3% of all prey consumed, followed by the small rice rats (22 g; 19.1%), Passeriformes (10–20 g; 14.3%) and tinamous (350–550 g; 9.5%). The average mean weight of mammalian prey (MWMP) was 287 g. Prey of 500 g appears to be important both numerically and in terms of biomass consumed by this small felid. Compared to other felines found in the Pampas, their diet had greater similarity to that of the jaguarundi (which also makes use of open environments), with lesser similarities to those of Geoffroy's cat and the margay [42]. These findings give support to the old reports that described the species preying mostly on cavies and ground birds [43].

5.2 Notes on habitat, activity and reproduction

The rare reports made by authors from the late 19th to the early 20th centuries already mentioned the species rarity and its "favored" habitat, the native tall

Figure 3.
Views of the Pampa cat's (Leopardus munoai) suitable habitat, its latrine (bottom middle) and tracks.

grasslands [44, 45]. Based on prey habits [42], it is possible to infer that Pampa cats prefer open fields with shrubby cover and edges of bogs to forage for terrestrial prey in the savannas of the Brazilian Pampas (**Figure 3**). Further support comes from the alternative ways to unlock the species habitat before telemetry information can properly elucidate this issue. A road-killed young male Pampa cat had a tick (*Amblyomma tigrinum*) whose life cycle is associated with weeds present only in virgin native fields. Casual observations lead us to assume that, although the species presents a greater diurnal activity, it would also present nocturnal activity (FD Mazim, pers. obs.), unlike the northern species of the country, *Leopardus braccatus*, whose activity is mostly diurnal [46, 47].

During four encounters with observations of females with cubs, the number of cubs varied between 1 and 2, with an average of 1.25. Birth estimates would be for early summer, late summer and early spring, and late spring/early summer (**Figure 4**). When they detect danger, the cubs enter armadillo burrows (*Euphractus sexcinctus*), or they crouch down to the ground, as do adults when hiding from an imminent threat.

6. The rarity of the species: Carnivores and the pampas

The summarization of the records obtained in the last 25 years was sufficient to obtain only 32 concrete records of the Pampa cat within the Brazilian Pampas. These comprised 2 (6.25%) camera-trap records, 9 (28%) visual observations, 17 (53%)

Figure 4.
Rare views, Pampa cat (Leopaurdus munoai) mother and cub playing (photo: Caio Sarmento Belleza) and a cub that was playing with a piece of tyre by the side of the highway (photo: Michel Correia).

road-kill specimens and 4 (12.5%) hunted individuals (**Table 1**). This would lead to an average of only 1.28 records per year. The overall relative abundance index (RAI) for all areas was a paltry 0.004 ind/100 trap-nights. Comparatively, rarity in terms of abundance of photographic records for Neotropical mammals was considered at 0.300 ind/100 trap-nights [63]. The inexorable rarity of *Leopardus munoai* is confirmed by comparing it with the observations of other sympatric cats, as well as other grassland carnivores for all indices – **Table 1**. Ironically, the Pampa cat is the rarest

Species	RAI-Pampa (ind./100 trap-nights)	Cam-trap records	Visual observations	Road-kills	Captures
Leopardus munoai Pampa cat	0.004	2 (0.001%)	9 (0.79%)	21 (0.89%)	0 (0%)
Leopardus geoffroyi Geoffroy's cat	0.442	221	34	88	103
Leopardus wiedii Margay	0.432	216	17	75	78
Herpailurus yagouaroundi Jaguarundi	0.024	12	6	18	4
Pseudalopex gymnocercus Pampas fox	1.430	715	422	935	473
Galictis cuja Lesser grisson	0.184	92	111	94	23
Conepatus chinga Molina's hog-nosed skunk	0.434	217	534	1103	67
Total		**1475**	**1133**	**2334**	**748**

RAI = relative abundance index.

Table 1.
Records of the Pampa cat (Leopardus munoai), other sympatric felids and carnivores in the Brazilian pampas between 1997 and 2022.

felid in the Pampas, 100 times rarer than Geoffroy's cat and even the margay, a species that is notorious for living in the forests though present in a mostly grassland biome.

At the Saicã site, two records were obtained after 1080 trap-nights or 25,920 hours of camera trapping that focused on the native fields free that have been from agriculture for at least 40 years. This would yield a relative abundance of 0.185 ind/100 trap-nights in the supposedly prime *L. munoai* habitat, which is far below the threshold of 0.300 ind/100 trap-nights proposed as the rarity index for medium-large Neotropical mammals on camera [48]. Considering the additional camera trapping carried out at other habitats not favored by the species (e.g., forest formations and crop edges), the relative abundance after 1975 trap-nights would drop to 0.101 ind./100 trap-nights, far below those presented by Geoffroy's cat (1.266 ind/100 trap-nights) and the margay (0.608 ind/100 trap-nights) in the same area. Nevertheless, *L. munoai*'s index was on par with that reported for its sister species, *Leopardus braccatus,* in the northern savannas (0.119 ind/100 trap-nights; [34]). Perhaps the Saicã estimate would be more appropriate in areas where the species might still be found.

In the 19th century, the Pampa cat was already considered rare by Ihering [44] and Araújo [45], both citing the species as being in population decline in the Coastal Plain and in the Serra do Sudeste (currently the municipalities of Pelotas, Canguçu, Morro Redondo, and Arroio do Padre), eastern-central portion of the Pampas, with a tendency to isolation in the native fields of the western border of the state of Rio Grande do Sul. Salvini [49] emphasized that in the fields of the extreme southeast of the Coastal Plain, between the municipalities of Santa Vitória do Palmar and Chuí, the species was practically extinct due to retaliatory and sport hunting by chicken farmers and the increasing reduction and degradation of native fields. In addition, it is a species that, since its type registration carried out in 1923, has less than 200 independent records throughout its entire geographical distribution zone [8, 17], which equates to less than two records per year, notably of individuals killed, run over, or hunted. The species became so scarce everywhere that any signs of it are becoming subject of publication [50].

7. Assessing the current species range and the expected population size

The model results were robust, with high predictive power (AUC = 0.89), low omission rate of test samples (o.r. = 0.13), and were highly significant ($P < 0.001$). The top three variables that influenced Pampa cat distribution were annual temperature range (% contribution = 44.6), mean annual rainfall (% contribution = 26.5), and isothermality (% contribution = 25.4). Suitability increased with mid values of annual temperature range (22–23°C, **Figure 5A**), rainfall (1350–1400 mm, **Figure 5B**), and increased isothermality (**Figure 5C**). Model results indicated a suitable area of 54,860 km^2 in 1997, which by 2021 had declined by 42%, resulting in 31,808 km^2 of suitable habitat left for the species in the Pampas region of the state (**Figure 6**). The bulk of this region is located in the central-southern portion of the state, along the Uruguayan border. Given the very few recent records of the species, it is likely that the Pampa cat is in fact absent from most of the suitable remaining habitat. On the other hand, if properly protected, the remaining suitable habitat is still large enough for reintroduction programs aided by ex situ conservation. Given the current rates of habitat loss in the Brazilian Pampas, it is expected that this suitable area will be even smaller as of 2023. Modeling conducted for Uruguay showed higher suitability for two main areas, both wetlands, while Tirelli's prediction indicated Rio Grande do Sul with most part of the suitable area remaining [17].

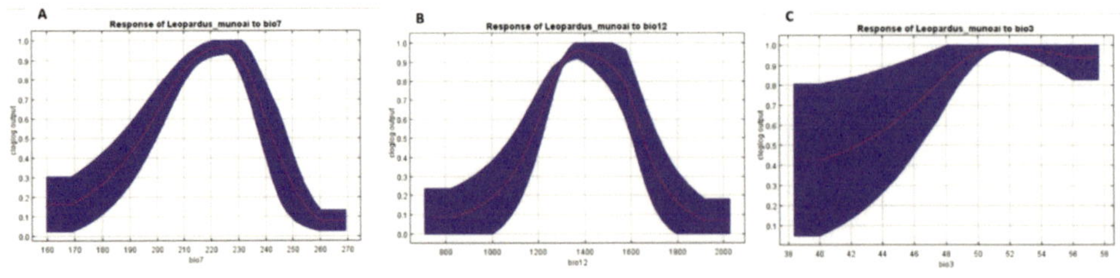

Figure 5.
Predicted habitat suitability of the Pampa cat (Leopardus munoai) in response to: A) annual temperature range; B) mean annual rainfall; C) isothermality.

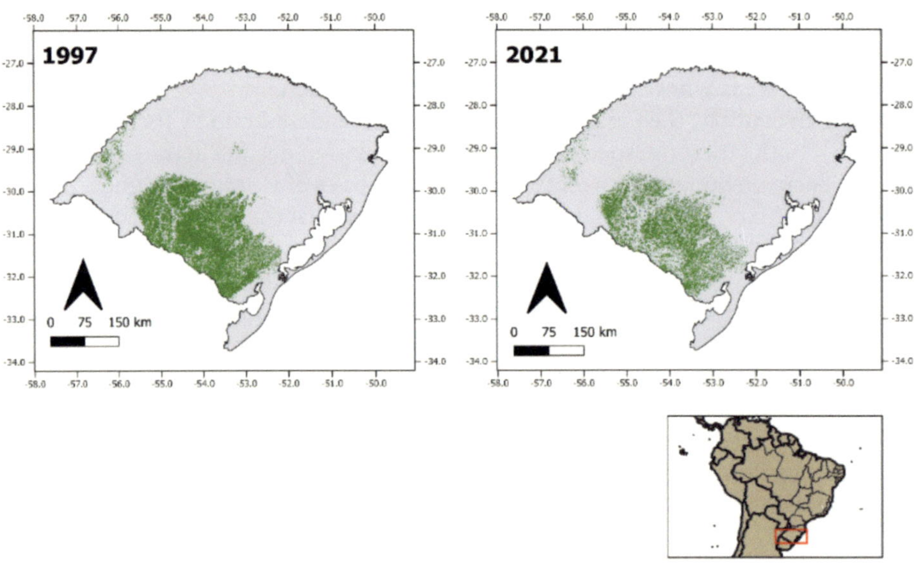

Figure 6.
Suitable habitat for Pampa cat in 1997 and 2021. Potential suitable habitat in 1997 was derived from Maxent models. Suitable habitat in 2021 was obtained by removing areas in which natural habitat was lost between 1997 and 2021.

Considering all the suitable areas remaining and the lower levels of population densities known for small Neotropical cats, 0.01 ind./km^2 and 0.0014 ind./km^2 [51], we would expect a total population of 45 to 318 individuals in Brazil. With a higher likelihood towards the lower values. The field-based best estimate of total population of no more than 50 individuals matched the lower predicted value. As such, we would consider the total remaining adult population of Pampa cat in Brazil to be 45–50 individuals. No field-based estimates and population information is available for Uruguay and Argentina. Bou et al. [52] estimated 144,483.3 km^2 of suitable habitat for the species in Uruguay, however with only 42 records in the country, it is likely that large swaths of otherwise suitable habitat are in fact devoid of Pampa cats.

From this conversion of the native matrix into agricultural areas, assuming that the same has happened in the registration points of Tirelli et al. [17], based on the 24 demographic scenarios evaluated, which, in various estimates, indicated that this feline might be critically endangered, the best-estimated scenario resulted in a population of 8866 mature individuals, which would decline to 1064, and which differs from the reality observed in the field, while the worst estimated scenario resulted in 62 mature individuals, currently reduced to eight mature specimens throughout its geographical distribution. Our current lower estimates of 45–50 individuals fall within similar levels of the worst scenario cited by Tirelli et al. [17]. Possibly relictual individuals survive as nomads, thus decreasing the chances of reproductive encounters and increasing the chances of being run over and exposure to predation by domestic dogs. To date, records of the species, especially visual observations and camera trapping, have never been repeated at the same point or location, at least in Brazil. Even in the five rare events in which individuals were observed with pups, they were never seen again in those same spots since these places had already been converted into soybean fields. Based on the difficulty involved in finding records, the rarity of herbaceous fields, the exponential conversion of native virgin fields into fields for crops, highway losses, hunting as a form of retaliation, predation by domestic dogs, possible contamination by diseases transmitted by domestic animals (dogs and cats) and possible problems of genetic stochasticity, it is believed that the most negative scenario is the most accurate. Not to mention that the conversion of native fields into soybean crops is widespread throughout the Brazilian Pampa, where arable fields are present, with deep and non-rocky soil. Considering the rarity of records and predictions for Uruguay [52], the projections of Tirelli et al. [17] and our field-based estimates, the total worldwide population of the Pampa cat would unlikely be much above 100–120 individuals, which is still very critical.

As an aggravating factor, there are no concrete records of the Pampa cat in conservation units in the Brazilian Pampas [53], and it is only mentioned that throughout its geographical distribution some protected areas have habitat suitability to support its presence [17]. At the same time, there is no allocation of government funds to create new protected areas nor intention to do so, which leaves populations at the mercy of the impacts or at the mercy of the management of private properties. However, within the current scenario, even if new conservation units are implemented, the disappearance and degradation of the fields leave the protected areas isolated. At the same time, since we do not know the minimum size of the species' territory and other basic aspects of spatial ecology and natural history, there is a risk that the size and place of implementation of a protected area are in regions that are not favorable for the maintenance of the species.

Moreover, another immediate negative factor, and a currently growing one in the rural matrix of the Pampas, is the passing of fields to the heirs of families that have traditionally been cattle breeders, and end up belonging to generations that settle in the urban area, and tend not to follow in the cultural footsteps of their ancestors in the gaucho cattle ranching tradition. These abandon traditional livestock farming and, after inheriting the fields, often opt to lease the land for the cultivation of rice and soybeans, since it is more profitable.

8. The dooming of an endemic species

There have been massive mammal extinctions in the Brazilian Pampas, such as those of the jaguar (*Panthera onca*), tapir (*Tapirus terrestris*), giant river otter (*Pteronura brasiliensis*), marsh deer (*Blastocerus dichotomus*), collared peccary (*Dicotyles tajacu*), white-lipped peccary (*Pecari tajacu*) and giant anteater (*Myrmecophaga tridactyla*). Additionally, there are only relictual populations of puma (*Puma concolor*), maned wolf (*Chrysocyon brachyurus*), and Pampas deer (*Ozotoceros bezoarticus*). All this coincides with the replacement by the agriculture of the traditional gaucho culture of cattle-raising in native fields [23, 44, 45, 49]. Thus, the Pampas is considered the Brazilian biome that loses more native vegetation than any other. At the same time, one in four hectares of native vegetation in the Pampas is secondary formation, i.e., one that has already been suppressed at least once and ended up regenerating [41].

The current agricultural exploitation, mainly the cultivation of soybeans and rice, has mischaracterized the native rural matrix of the Pampas in Brazil and also in the Uruguayan Pampa (natural formations forming the Uruguayan savanna), thus restricting continuously and annually the availability of favorable fields for the occurrence of the Pampa cat. This fact seems to be the main motivator of the records of the species along the highways, even the visual ones, since in the vast majority of cases, the domain bands (linear land space, present between the highway and private properties) are not suppressed and are impacted by grain plantations. This condition allows the development of vegetation that is typical of native fields and the presence of their natural prey, which becomes the only fields that are preserved with aspects close to the original configuration in most regions of the Pampas. At the same time that the domain lanes of the highways offer conditions for the maintenance of the population of the Pampa cat, they result in direct mortality due to road accidents, since it brings individuals closer to traffic and restricts them from living near vehicle traffic. Another historical and cultural impact on the Pampa cat and other felines of the Pampas refers to hunting motivated by prevention or retaliation, as a way of controlling alleged predators of domestic birds [53, 54].

It is crucial to emphasize that most of the Uruguayan savanna ecoregion is categorized as having the greatest future threat of conversion, with virtually all acreage severely altered by agriculture [55]. As an example, between 2002 and 2016, a 190-kilometre stretch of highway (BR-293), located between the cities of Pinheiro Machado and Dom Pedrito, in the extreme south of the Pampa biome, recorded 11 individuals run over and, as of 2017, this repetition or concentration was no longer evidenced, since two individuals were counted. Notably, this occurred in a period when native fields with livestock had accelerated replacement by soybean crops [41].

In Brazil, the Pampa cat lives in the smallest and most-degraded ecosystem and biome, that also has the greatest disregard in terms of the release of government resources for surveillance and research. This is perhaps motivated by the fact that the Pampas biome does not have the environmental lobby and charisma that the Amazon, the Atlantic Forest and the Pantanal have, both nationally and, above all, globally. This scenario also reduces the chances of financing via private initiatives. The natural absence of large forest cover (showcase of conservation) makes the Pampas biome (as well as the Uruguayan savanna) not very emblematic, added to the absence of charismatic species such as the jaguar as well as other large mammals that became extinct in the 19th century [44, 56, 57].

Since 2008, the Pampa cat is no longer recorded in the southern and central portions of the Coastal Plain of the Pampas biome (state of Rio Grande do Sul). However, before this period, only two confirmed records were obtained in this region, both on the edges of the BR-116, one in 2002 from a visual observation (municipality of Cristal) and another in 2008 from an individual found run over (municipality of Jaguarão). It is in this region of the extreme east of the Pampas biome that the Patos/Mirim lagoon complex is located, which has the phytophysionomy of a floodplain and water-logged fields, with natural environments converted almost entirely into rice crops since the 1940s, though with greater intensity from the 1970s onwards. There the shallower water-logged fields were drained for rice planting, which, as of 2010, has been replaced by soybean crops.

It is important to highlight that, since 2010, the two main highways (BR-101 and BR-116) that cover 100% of the surface of the Pampas Coastal Plain (north-south direction) are monitored daily by expert consultants and traffic operators of the highway concessionaires in the search for run over specimens, as condition of the environmental license that was granted. However, even with this daily effort, the Pampa cat has not been recorded.

Another scenario that is capable of revealing the rarity of the species is in the official records of the supervisory agencies. In all the records, only two individuals of Pampa cat have been admitted to fauna rehabilitation centres in Brazil, and both were orphaned cubs, one male specimen in the early 2000s (municipal zoo of Cachoeira do Sul) and the other a female in 2014 (CETAS/IBAMA/RS and then sent to the zoo at Gramado). In Uruguay, three individuals are currently part of an ex situ conservation program, as part of the M'bopicuá BioPark, with the aggravating factor that two specimens are siblings, thus reducing opportunities for non-parental crosses, in other words, the species may become rapidly and concomitantly extinct in nature and on the planet.

The doomsday scenario presented, although referring to Brazil, in great part would also reflect the remaining areas of its range, and the little information known does not differ from what is presented here. As such, the road to extinction that is portrayed for the Pampa cat in Brazil likely reflects its overall scenario [17, 52]. The last extinction of a felid species are known only from the time of the Pleistocene, when sabre-tooth tigers, cave-lions, the American cheetah and others disappeared [58]. There have been unfortunate extinctions of several large felids, such as the South China, Java and Bali tigers, and the Atlas lion, but these refer to subspecies/populations and not of an entire species. There are certainly other small and large felid species and subspecies on the road to extinction, such as the Iberian lynx (*Lynx pardinus*), the tiger (*Panthera tigris*), the Borneo bay cat (*Catopuma badia*), the flat-headed cat (*Prionailurus planiceps*), and the Andean cat (*Leopardus jacobita*), to name a few [59–62], but none of them are remotely close to the situation portrayed here for the Pampa cat. The Iberian lynx was the species that got the closest to becoming extinct, but after a massive effort to protect and expand its population, the species recovered from a few 100 or so to nearly 400 individuals and is currently on the rise to the point of having been downlisted from critically endangered to endangered [59]. Unfortunately, the Pampa cat is in a worse situation than that of the Iberian lynx in the late 1970s. There is no known population established anywhere (not to mention in protected areas), and no strong financial support, particularly because the species has not been even properly recognized as such by the IUCN, nor does it have strong financial support from the European Union that the Iberian lynx had. Without taking immediate and forceful action, as the old saying goes, "all roads lead to Rome", the road will soon lead to extinction, the first one of a full species in the modern age!

This sad journey should serve as a warning lesson for two other small felids from Brazil that are also found in agricultural lands and that have already suffered considerable habitat losses, so far being pushed along the same path towards extinction as the Pampa cat. These are the Brazilian Pampa cat (*Leopardus braccatus*), its sister species, and the savanna tiger cat (*Leopardus tigrinus*), both found in the dwindling savannas of Brazil. There is still time for these two; however, in order to avoid the demise of the remaining Pampa cats, a conservation action plan needs to be put into place immediately.

9. Conservation action plan for immediate implementation

The main conservation action for the Pampa cat should be the protection of its habitat, the native herbaceous fields. However, for the protection of these fields, management is necessary, which fundamentally includes the presence of extensive grazing (cattle or sheep) without confinement. From the moment the cattle are removed from the fields, the configuration of the field vegetation becomes shrubby. On the other hand, with large stockings or confinement, the configuration of the fields is reduced to a thin carpet of grasses. Therefore, a field without livestock or with a high stocking of cattle tends to result in the absence of the Pampa cat.

As a crucial measure to maintain the extensive form of breeding, there is an urgent need to create and implement environmental quality seals that value animal protein (green beef) produced extensively in virgin native fields or those in an advanced stage of recovery. This would be an alternative to add value to livestock, encouraging the rural producer not to migrate and convert the fields into soybean crops. The scope of these programs to monetize the protein produced in the native fields should cover the cross-border scope between Argentina, Brazil, and Uruguay, especially in the international corridors of more connected natural areas, as suggested by Tirelli et al. [17].

The creation of the seal of quality must include the name of the Pampa cat. Given that the species occurs in an environment intended for food production, its presence should leverage monetization; otherwise, it may be viewed negatively, and accelerate the process of converting fields into crops, since landowners, upon becoming aware of the rarity of this species, may associate and fear that the presence of the Pampa cat on their properties will result in the embargo of their fields. In summary, the presence of this feline must be seen as a possibility of increasing income, in other words, making livestock economically competitive with purely agricultural activities (soybean and rice crops).

A certification program for soybean and rice crops also appears as an alternative arrangement, especially in the subsidy of agricultural areas with the planning of field corridors between crops, thus subsidizing the areas with the presence of the Pampa cat.

- Program 1: Create and implement a monetization program (valorization $) of animal protein that comes from livestock raised in native fields (green beef).

- Program 2: Create and implement a program to monetize (valorization $) soybeans and rice by decreasing planted areas while preserving native field corridors between properties and highway domain strips.

- Program 3: Create and implement an ex situ breeding program of the Pampa cat (increase genetic variability) in different institutions, as well as zoos, conservation centres and private properties, considering the risk of tragedies from fire, flood, and theft, among others.

    - Subprogram 3.1. Capture and take free-living individuals to centres included in the ex situ conservation program.

- Program 4: New census and refinement of possible relictual populations or individuals with the established territory.

- Program 5: Characterize the basic needs of the species (spatial ecology, habitat use, activity, genetics, zoonoses, and reproduction).

- Program 6: Set up a consultancy service with exclusive dedication to the execution of the tasks of the action plan.

    - Subprogram 6.1: Set up a multidisciplinary consultancy team (professional feline specialist, wildlife veterinarian, agronomist and professionals in the field of communication, marketing, economics, and culture).

    - Subprogram 6.2: Multidisciplinary consultancy service working in an integrated way with the relevant agencies (IBAMA, State Secretariat, Public Prosecutor's Office, Environmental Military Police).

    - Subprogram 6.3: Multidisciplinary consultancy service working in an integrated way with parliamentarians linked to agricultural cooperatives to improve relations between rural producers, researchers, and public managers.

10. Concluding remarks

Based on the exponential conversion of native fields, initially into rice plantations, and currently mostly in soybean crops, and the relation to the annual reduction in records, we can infer that the Pampa cat will not survive extinction per se and require immediate human intervention. Unfortunately, measures such as environmental education do not offer guarantees of changes in the current situation – which are necessary after the start of the conservation action plan – because the Pampa cat lives in an environment that is destined to become a voluminous and profitable economic area of production (the fields) and, if its presence in native fields with extensive livestock is not a clearly advantageous income alternative, as or more profitable than soybeans, its preferred habitat will inevitably be annihilated.

Associated with exponential habitat loss, another main factor in the struggle of the Pampa cat against imminent extinction, is its specialist ecology in terms of quality and physiognomies of the fields, where, although its concrete bioecological needs and its basic spatial ecology are still unknown, the records of two decades demonstrate that the species depends on specific typologies of native fields, especially virgin fields, without shrubby vegetation or simply summarized as sparse grassy fields. In a preliminary way, its records have an association with virgin herbaceous native fields, composed of weeds in the dry areas and straw in the humid areas, both with moderate grazing (livestock fields), since the withdrawal of cattle grazing makes the fields shrubby, and thus results in the disappearance of this feline. It is important to report that, in fields without cattle breeding, where shrubby vegetation is formed by grazing, the species was not recorded. The same is observed in fields with large numbers of cattle, which are restricted to a sparse configuration composed of grasses.

Therefore, it is recommended that the actions proposed as an attempt to reduce the imminent risk of extinction of the species should be used concomitantly (research and conservation), with the main focus on the monetization of livestock in the native fields, through the creation of environmental seals. This guarantees the presence of physical spaces (fields) that are favorable to the species, in order to be seen as a feasible guarantee of income by landowners; otherwise, there will be no justification for maintaining the fields, given the great profitability generated by soybean crops. Another consequence of the loss of native fields is the annihilation not only of the Pampa cat and the biodiversity associated with it, but also of an intangible anthropological heritage, notably the culture of the gaucho people, which was forged by cattle breeding.

The conservation of native fields for livestock is the most sustainable campesino economic system for the conservation of rural biodiversity and, fundamentally, for the regional culture (gaucho), since livestock farming in the Pampas does not result from deforestation, and, since its beginning around 1620 [63], it has been conducted on natural pastures. In summary, this is the most appropriate economic vocation of the Pampas and is the most sustainable and responsible for maintaining regional biodiversity and culture. Unfortunately, the current scenario reveals that new farmers or field heirs tend not to continue the socioeconomic practices conducted by their livestock farming ancestors, and they are inclined to convert native fields into crops in order to achieve greater financial returns.

The negative factors regarding the Pampa cat have been reported for at least a century and indicate an increase in this pressure every year, which restricts *Leopardus munoai* to a species composed of relictual and perhaps nomadic individuals, possibly with inbreeding occurring. In this way, the current scenario allows us to postulate that the Pampa cat is in fact the most threatened feline on Earth and may become extinct without even being formally recognized and known. Unfortunately, if the current scenario is maintained, this endemic feline of the Pampas will be extinct in the wild in less than a decade.

Author details

Fábio Dias Mazim[1,2], Paulo Guilherme Carniel Wagner[3,4], Lester Alexander Fox-Rosales[5], Alisson da Rosa Boÿink[6] and Tadeu Gomes de Oliveira[1,7]*

1 Instituto Pró-Carnívoros, Atibaia, São Paulo, Brazil

2 Ka'aguy Consultoria Ambiental Ltda, Pelotas, Rio Grande do Sul, Brazil

3 Centro de Triagem de Animais Silvestres, Superintendência do Ibama no Rio Grande do Sul, Brazil

4 Laboratório de Protozoologia e Rickettsioses Vetoriais, Faculdade de Veterinária-Universidade Federal do Rio Grande do Sul, Porto Alegre, Brazil

5 Programa de Pós-Graduação em Ciência Animal da Universidade Estadual do Maranhão, São Luís, Maranhão, Brazil

6 Zoológico Municipal de Cachoeira do Sul, Rio Grande do Sul, Brazil

7 Departamento de Biologia, Universidade Estadual do Maranhão, Campus Universitário Paulo VI, São Luís, MA, Brazil

*Address all correspondence to: tadeu4@yahoo.com

References

[1] Oliveira TG, Cassaro K. Guia de Campo dos Felinos do Brasil. São Paulo: Instituto Pró-Carnívoros/Fundação Parque Zoológico de São Paulo/SZB/Pró-Vida Brasil; 2005

[2] Kitchener AC, Breitenmoser-Würsten C, Eizirik E, Gentry A, Werdelin L, Wilting A, et al. A revised taxonomy of the Felidae. The final report of the cat classification task force of the IUCN/SSC cat specialist group. Cat News Special Issue. 2017;(11):80

[3] Nascimento FO, Cheng J, Feijó A. Taxonomic revision of the pampas cat Leopardus colocola complex (Carnivora: Felidae): an integrative approach. Zoological Journal of the Linnean Society [Internet]. 28 May 2020;**191**(2):1-37. Available from: https://academic.oup.com/zoolinnean/advance-article/doi/10.1093/zoolinnean/zlaa043/5848237

[4] Olson DM, Dinerstein E, Wikramanayake ED, Burgess ND, Powell GVN, Underwood EC, et al. Terrestrial ecoregions of the world: A new map of life on earth. Bioscience. 2001;**51**(11):933-938

[5] Johnson WE, Slattery JP, Eizirik E, Kim JH, Menotti Raymond M, Bonacic C, et al. Disparate phylogeographic patterns of molecular genetic variation in four closely related south American small cat species. Molecular Ecology. 1999;**8**(s1):S79-S94. [Internet] Available from: http://doi.wiley.com/10.1046/j.1365-294X.1999.00796.x

[6] Sartor CC. Estrutura populacional de Leopardus colocolo (Carnivora, Felidae): definindo unidades demográficas para a conservação da espécie no Brasil [Internet]. Porto Alegre: Universidade Federal do Rio Grande do Sul; 2016 Available from: https://lume.ufrgs.br/handle/10183/143875

[7] da Silva SA, Trigo TC, de Oliveira TG, Silveira L, Eizirik E. Phylogeographic analyses of the pampas cat (Leopardus colocola; Carnivora, Felidae) reveal a complex demographic history. Genetics Molecular. 2018;**41**(1 suppl 1):273-287. [Internet] Available from: http://www.scielo.br/scielo.php?script=sci_arttext&pid=S1415-47572018000200273&lng=en&tlng=en

[8] Martínez-Lanfranco JA, González EM. The oldest available name for the pampas cat of the Uruguayan Savannah ecoregion is Leopardus fasciatus (Larrañaga 1923). Therya. 2022;**13**(3):259-264. [Internet] Available from: http://132.248.10.25/therya/index.php/THERYA/article/view/1187

[9] García-Perea R. The pampas cat Group (genus Lynchailurus Severtzov, 1858) (Carnivora: Felidae), a systematic and biogeographic review. American Museum Novitates. 1994;**3096**:35

[10] Overbeck G, Muller S, Fidelis A, Pfadenhauer J, Pillar V, Blanco C, et al. Brazil's neglected biome: The south Brazilian Campos. Perspectives in Plant Ecology, Evolution and Systematics. 2007;**9**(2):101-116. [Internet] Available from: https://linkinghub.elsevier.com/retrieve/pii/S1433831907000303

[11] Andrade BO, Bonilha CL, Overbeck GE, Vélez-Martin E, Rolim RG, Bordignon SAL, et al. Classification of south Brazilian grasslands: Implications for conservation. Jansen F, editor. Applied Vegetation Science. 2019;**22**(1):168-184. [Internet] Available from: https://onlinelibrary.wiley.com/doi/10.1111/avsc.12413

[12] Menezes LS, Ely CV, Lucas DB, Minvervini-Silva GH, Vélez-Martin E, Hasenack H, et al. Reference values and drivers of diversity for south Brazilian grassland plant communities. Anais da Academia Brasileira de Ciências. 2022;**94**(1):1-20. [Internet] Available from: http://www.scielo.br/scielo.php?script=sci_arttext&pid=S0001-37652022000101021&tlng=en

[13] Loyola RD, Kubota U, da Fonseca GAB, Lewinsohn TM. Key Neotropical ecoregions for conservation of terrestrial vertebrates. Biodiversity and Conservation. 2009;**18**(8):2017-2031 [Internet] Available from: http://link.springer.com/10.1007/s10531-008-9570-6

[14] FZB. Decreto N 51.797 de 8 de Setembro de 2014. Porto Alegre: Estado do Rio Grande do Sul Assambleia Legislativa; 2014

[15] GonzálezEM,Martínez-LanfrancoJA, Juri E, Rodales AL, Botto G, Soutullo Á. Mamíferos. In: Soutullo A, Clavijo C, Martínez-Lanfranco JA, editors. Especies prioritárias para la conservación en Uruguay. Montevideo: SNAP/DINAMA/MVOTMA/DICYT/MEC; 2013. pp. 175-207

[16] Lucherini M, Eizirik E, de Oliveira TG, Pereira JA, Williams RSR. *Leopardus colocolo* [Internet]. The IUCN Red List of Threatened Species 2016: e.T15309A9720444. 2015. Available from: https://www.iucnredlist.org/species/15309/97204446

[17] Tirelli FP, Trigo TC, Queirolo D, Kasper CB, Bou N, Peters F, et al. High extinction risk and limited habitat connectivity of Muñoa's pampas cat, an endemic felid of the Uruguayan Savanna ecoregion. Journal for Nature Conservation. 2021;**62**(May):126009. [Internet] Available from: https://linkinghub.elsevier.com/retrieve/pii/S161713812100056X

[18] de Almeida CE, D'Elía G, Percequillo AR. Taxonomic review of genus Sooretamys Weksler, Percequillo & Voss (Rodentia: Cricetidae: Sigmodontinae): An integrative approach. Zoological Journal of the Linnean Society. 2014;**171**(4):842-877. [Internet] Available from: https://academic.oup.com/zoolinnean/article-lookup/doi/10.1111/zoj.12146

[19] Feijó A, Cordeiro-Estrela P. Taxonomic revision of the Dasypus kappleri complex, with revalidations of Dasypus pastasae (Thomas, 1901) and Dasypus beniensis Lönnberg, 1942 (Cingulata, Dasypodidae). Zootaxa. 2016;**4170**(2):271. [Internet] Available from: http://biotaxa.org/Zootaxa/article/view/zootaxa.4170.2.3

[20] Feijó A, Patterson BD, Cordeiro-Estrela P. Taxonomic revision of the long-nosed armadillos, genus Dasypus Linnaeus, 1758 (Mammalia, Cingulata). Hubbe A, editor. PLoS One; 2018;**13**(4):e0195084. [Internet] Available from: https://dx.plos.org/10.1371/journal.pone.0195084

[21] Martino D. Conservación de praderas en el conosur: valoración de las áreas protegidas existentes. Revista Ecosistemas. 2004;**13**(2):1-12

[22] Overbeck GE, Vélez-Martin E, Scarano FR, Lewinsohn TM, Fonseca CR, Meyer ST, et al. Conservation in Brazil needs to include non-forest ecosystems. Loyola R, editor. Diversity and Distributions. 2015;**21**(12):1455-1460. [Internet] Available from: https://onlinelibrary.wiley.com/doi/10.1111/ddi.12380

[23] Queirolo D. Diversidade e padrões de distribuição de mamíferos dos campos

do uruguai e sul do brasil. Boletín de la Sociedad Zoológica del Uruguay. 2016;**25**(2):92-247. [Internet] Available from: http://szu.org.uy/journal/index.php/Bol_SZU/article/view/11

[24] Ximénez A. Nueva subespecie del gato pajero en el Uruguay Felis colocola muñoai n. ssp. Comunicaciones zoológicas del Museo de Historia Natural de Montevideo. 1961;**5**:1-8

[25] Larrañaga DA. Escritos de Don Dámaso Antonio Larrañaga, tomo II. Montevideo: Instituto Histórico y Geográfico del Uruguay; 1923

[26] IBGE. Mapa da vegetação do Brasil e Mapa de Biomas do Brasil [Internet]. 2004. Available from: https://ww2.ibge.gov.br/home/presidencia/noticias/21052004biomashtml.shtm#sub_download.2004.

[27] Peel MC, Finlayson BL, McMahon TA. Updated world map of the Köppen-Geiger climate classification. Hydrol earth. Systems Science. 2007;**11**(5):1633-1644. [Internet] Available from: https://hess.copernicus.org/articles/11/1633/2007/

[28] Porto ML. Os campos sulinos: sustentabilidade e manejo. Cienc Ambient. 2002;**24**(4):119-138

[29] Buriol GA, Estefanel V, de Chagas ÁC , Eberhardt D. Clima e vegetação natural do estado do Rio Grande do Sul segundo o diagrama climático de Walter e Lieth. Ciência Florest. 2007;**17**(2):91-100. [Internet] Available from: https://periodicos.ufsm.br/cienciaflorestal/article/view/1940

[30] Quadros FL, de Pillar VP. Transições florestas-campo no Rio Grande do Sul. Cienc Ambient. 2002;**2**:109-118

[31] Marchiori JNC. Fitogeografia do Rio Grande do Sul: campos sulinos. Porto Alegre: Edições ES; 2004. p. 110

[32] Verdum R. O pampa: ainda desconhecido. Revista do Instituto Humanitas Unisinos. 2006;**183**:4-9

[33] Marques RV, Mazim FD. A utilização de armadilhas fotográficas para o estudo de mamíferos de médio e grande porte. Cad La Salle XI. 2005;**2**(1):219-228

[34] de Oliveira TG, Lima BC, Fox-Rosales L, Pereira RS, Pontes-Araújo E, de Sousa AL. A refined population and conservation assessment of the elusive and endangered northern tiger cat (Leopardus tigrinus) in its key worldwide conservation area in Brazil. Global Ecology and Conservation. 2020;**22**:1-14

[35] Fox-Rosales LA, de Oliveira TG. Habitat use patterns and conservation of small carnivores in a human-dominated landscape of the semiarid Caatinga in Brazil. Mammalian Biology. 2022;**102**(2):465-475. [Internet] Available from:. DOI: 10.1007/s42991-022-00245-3

[36] Fox-Rosales LA, de Oliveira TG. Interspecific patterns of small cats in an intraguild-killer free area of the threatened Caatinga drylands, Brazil. Cristescu B, editor. PLoS One. 2023;**18**(4):e0284850. [Internet] Available from:. DOI: 10.1371/journal.pone.0284850

[37] Sikes RS, Gannon WL. Guidelines of the American Society of Mammalogists for the use of wild mammals in research. Journal of Mammalogy. 2011;**92**(1):235-253. [Internet] Available from: https://academic.oup.com/jmammal/article-lookup/doi/10.1644/10-MAMM-F-355.1

[38] O'Brien TG, Kinnaird MF, Wibisono HT. Crouching tigers, hidden prey: Sumatran tiger and prey populations in a tropical forest landscape. Animal Conservation. 2003;**6**(2):131-139

[39] Phillips SJ, Anderson RP, Schapire RE. Maximum entropy modeling of species geographic distributions. Ecological Modelling. 2006;**190**(3-4):231-259. [Internet] Available from: https://linkinghub.elsevier.com/retrieve/pii/S030438000500267X

[40] Fick SE, Hijmans RJ. WorldClim 2: New 1-km spatial resolution climate surfaces for global land areas. International Journal of Climatology. 2017;**37**(12):4302-4315. [Internet] Available from: https://onlinelibrary.wiley.com/doi/10.1002/joc.5086

[41] Souza CM, Shimbo ZJ, Rosa MR, Parente LL, Alencar A, BFT R, et al. Reconstructing three decades of land use and land cover changes in Brazilian biomes with Landsat archive and earth engine. Remote Sensing. 2020;**12**(17):2735. [Internet] Available from: https://www.mdpi.com/2072-4292/12/17/2735

[42] Migliorini RP, Peters FB, Favarini MO, Kasper CB. Trophic ecology of sympatric small cats in the Brazilian Pampa. PLoS One. 2018;**13**(7):1-13

[43] Cabrera A, Yepes J. Mamíferos Sudamericanos. Buenos Aires: EDIAR; 1961

[44] Ihering HV. Os mamíferos do Rio Grande do Sul. In: Gundlach K, editor. Annuário do Estado do Rio Grande do Sul para o Anno 1892. Porto Alegre. 1892. pp. 96-123

[45] Araujo F. Principais representantes da fauna do Município - Mamífero s e Aves. Pelotas: Ed. Livraria Comercial; 1897. pp. 24-27

[46] de Oliveira TG, Tortato MA, Silveira L, Kasper CB, Mazim FD, Lucherini M, et al. Ocelot ecology and its effect on the small- felid guild in the lowland neotropics ocelot ecology and its effect on the. In: Macdonald DW, Loveridge AJ, editors. Biology and Conservation of Wild Felids. 1st ed. New York: Oxford University Press; 2010. pp. 559-580

[47] Lima BC. Of Small Cats and Dogs: Interspecific Relationships of Wild Cats and Domestic Carnivores in the Northern Savannas of Brazil. São Luís: Maranhão State University; 2020

[48] de Oliveira TG, Michalski F, Botelho ALM, Michalski LJ, Calouro AM, Desbiez ALJ. How rare is rare? Quantifying and assessing the rarity of the bush dog Speothos venaticus across the Amazon and other biomes. Oryx. DOI: 10.1017/S0030605316000624

[49] Salvini RO. Pindurama Sulina: O que produz e o que deveria produzir o município de Santa Vitória. Rio Grande, Brasil, Professor Lyceu Salsiano Leão XIII. 1936. p. 104

[50] Álvarez Pinna CM, López MJ, Pereira JA. Nuevo registro del gato de pajonal uruguayo Leopardus munoai (Ximénez, 1961) (Mammalia: Carnivora: Felidae) en el centro de la República Oriental del Uruguay. Notas sobre Mamíferos Sudamericanos. 2023;**05**(1):001-004. [Internet] Available from: https://ojs.sarem.org.ar/index.php/nms/article/view/907

[51] de Oliveira TG, Mazim FD, Marques RV, Fox-Rosales LA, Kasper CB, Marinho PH, et al. Demographic and spatial patterns of small felids in Brazil and the ocelot effect. In: 100th Annual Meeting of the American Society of Mammalogists. Online. Even3Publicações; 2021

[52] Bou N, Cuyckens GAE, González EM, Meneghel M. Conservation planning

in Uruguay based on small felids (Leopardus spp.) as umbrella species. Studies on Neotropical Fauna and Environment. 2019;**54**(3):169-180. [Internet] Available from:. DOI: 10.1080/01650521.2019.1669421

[53] Mazim FD, Dias RA, Fermun M, Salazar EA, Silva JCF. Novos registros de carnívoros para o sul do Rio Grande do Sul. In: Congresso Brasileiro de Mastozoologia. Belo Horizonte: Suprema; 2003. p. 47

[54] Peters F, Mazim FD, Favarini MO, Soares JBG, de Oliveira TG. Caça preventiva ou retaliativa de felinos por humanos no extremo sul do Brasil. In: Lasso CA, Hoogesteijn R, Díaz-Pulido A, Payán E, Castaño-Uribe C, editors. Conflictos entre felinos y humanos en América Latina. Bogotá, Colombia: Instituto de Investigación de Recursos Biológicos Alexander von Humboldt; 2016. p. 311-321

[55] Oakleaf JR, Kennedy CM, Baruch-Mordo S, West PC, Gerber JS, Jarvis L, et al. A world at risk: Aggregating development trends to forecast global habitat conversion. Kumar L, editor. PLoS One. 2015;**10**(10):e0138334. [Internet] Available from: https://dx.plos.org/10.1371/journal.pone.0138334

[56] Indrusiak C, Eizirik E. Carnívoros. In: Bencke GA, Reis RE, editors. Livro vermelho da fauna ameaçada de extinção no Rio Grande do Sul. Porto Alegre: Edipucrs; 2003. p. 632

[57] Mahler J, Schneider JKF. Ungulados. In: Livro Vermelho da Fauna Ameaçada de Extinção no Rio Grande do Sul. Porto Alegre: Edipucrs; 2003. pp. 247-565

[58] Antón M, Clark HF. The Big Cats & Their Fossil Relatives: An Illustrated Guide to their Evolution and Natural History. 1st ed. New York: Columbia University Press; 1997

[59] Rodríguez A, Calzada J. *Lynx pardinus*. The IUCN Red List of Threatened Species. 2020:e.T12520A174111773. DOI: 10.2305/IUCN.UK.2015-2.RLTS.T12520A174111773.en

[60] Hearn A, Brodie J, Cheyne S, Loken B, Ross J, Wilting A. *Catopuma badia*. The IUCN Red List of Threatened Species. 2016:e.T4037A112910221. DOI: 10.2305/IUCN.UK.2016-1.RLTS.T4037A50650716.en

[61] Villalba L, Lucherini M, Walker S, Lagos N, Cossios D, Bennett M, et al. *Leopardus jacobita*. The IUCN Red List of Threatened Species. 2016:e.T15452A50657407. DOI: 10.2305/IUCN.UK.2016-1.RLTS.T15452A50657407.en

[62] Goodrich J, Wibisono H, Miquelle D, Lynam AJ, Sanderson E, Chapman S, et al. Panthera tigris. The IUCN Red List of Threatened Species. 2022:e.T15955A214862019. DOI: 10.2305/IUCN.UK.2022-1.RLTS.T15955A214862019.en

[63] Flores M. Historia do Rio Grande do Sul. 5th ed. Porto Alegre: Nova Dimensão; 1996

Chapter 3

Reviving Chinese Pangolin Conversation: A Brief Knowledge Sharing

Suman Acharya

Abstract

Chinese pangolin is a unique critically endangered small mammal that is covered by scales and feed on termites and ants. It is dark brown or dark gray in color and lives in coniferous and broad-leaf forest, agricultural land, barren land, bamboo forest, grassland Chinese pangolin is widely distributed in the continent of Asia but the detailed studies on population status, ecology, behavior, and illegal trade is lacking. With the developmental activities escalated in Asian countries, the threats to the pangolin population and potential habitats are increasing. Additional athropogenic factors such as deforestation, encroachment, forest fire, over exploitation, hunting, poaching, and illegal trade have caused steep decline in the number of Chinese pangolin in wild. In fact, the pangolin is the most illegally traded animal in the contemporary world. Therefore, it is crucial that we revive and prolong communication regarding the current global and national status, distribution, behavior, illegal trade, and conservation practices across scales to promote the long-term conservation of the Chinese pangolin population and habitat. In the end, this chapter provides some important policy recommendations to promote Chinese pangolin conversation.

Keywords: Chinese pangolin, critically endangered, distribution, threats, conservation

1. Introduction

Pangolin is a small mammal that is widely known as a scaly anteater because its body is covered with a tough overlapping scales that is made up of keratin. It mostly feeds on ants and termites using an extraordinarily long and sticky tongue, and in the case of threatened is able to roll itself up into a tight ball quickly [1]. There are Eight species of pangolin distributed globally. The four species are categorized as Asian pangolin and four are African pangolin. The global population of these species of pangolin are unknown whereas under the IUCN Red List of Threatened Species all eight species are marked as globally threatened. Furthermore, the IUCN Red List has declared that the population of all eight species of pangolin is in decreasing trend throughout the world [2]. All species of pangolin are listed in Appendix I in the Convention on International Trade in Endangered Species of Wild Fauna and Flora [3]. Pangolin is shy, non-aggressive, solitary and burrowing strange mammal that is protected nationally as well as internationally but their biological details are unknown to stakeholders. This chapter reviews the taxonomy, ecology, habitat, population, distribution, conservation status and threats of Chinese pangolin *(Manis pentadactyla)* in detail and provides some recommendations in Chinese pangolin studies.

2. Taxonomy

Based on the morphological and genetic evidences, Chinese pangolin is specified under the genus "Manis" [4]. The name "*Manis pentadactyla*" is derived from two Latin words, '*manes*' refers to the nocturnal behavior and unusual appearance of pangolin and '*pentadactyla*' refers to five fingers on both fore and hind limbs. In Roman belief, the pangolin is known as the spirit of the ghost because of their nocturnal movement [5, 6]. Chinese pangolin is still considered as the sign of bad luck in different parts of the world including Nepal.

3. Ecology

Closely observing the ecological behavior of Chinese pangolin is exceptionally tough because of its lower population throughout the home-ranges and its nocturnal, secretive, shyness, and elusive nature. Therefore, most of the ecological description are based from the researches done in China and Taiwan where the population of the species is comparatively higher than other countries [6].

All species of pangolin including Chinese pangolin sleeps in hollows and logs during the day time and emerges in the evening to forage for ants and termites [1]. Chinese pangolin appears exceptionally different than other mammals. Some of the unique body appearances including large ear pinna, a post-anal depression in the skin and a narrowing near the distal end of the tail distinguish Chinese pangolin from other Asian pangolin [7]. As a defense, Chinese pangolin rolls into a ball and during these configuration soft tissues are protected or hidden by scales [8]. Its fore limbs are powerful and contain large digging claws, while hind limbs are shorter and contain smaller claws. It has a thick and long tail covered with large (40 cm long and 2–5 cm diameter), round overlapping scales formed from fused hair that is dorsally rounded and ventrally flattened, prehensile and very muscular [6]. Its tail is shorter as compared to other pangolin species and is less than half of the total body length [9]. The body of the species is entirely covered with the scales made up of keratin that grown from the body skin forming a grid that grow continuously [10]. The purpose of the scales is to cover and save the surface of head, trunk, lateral surfaces of trunk, limbs, and tail from external injuries and attacks. The color of Chinese pangolin is usually dark brown or dark gray (**Figure 1**) [10]. Whereas, the facial skin is yellow-pinkish gray [7].

Male Chinese pangolin is larger than female ones. The mass of Chine pangolin ranges from 2.35 kg (young, sexually matured female) to 7.0 kg (fat male), similarly body length ranges from 545 mm (young female) and 795 mm (male) [8]. However, some Chinese pangolin weight exceptionally heavy (i.e. more than 8 kg) [6]. Chinese pangolin usually digs two different kinds of burrows based on their use. The resident burrows are used for resting and sleeping purpose and the feeding burrows are used for feeding on ants and termites existing under the earth's surface [12]. Usually it uses the feeding burrow for a shorter period of time as compared to resident burrows

Figure 1.
Chinese pangolin. Source: Tulsi Laxmi Suwal [11].

where they spend longer time for resting and sleeping purposes. The Chinese pangolin prefer sunlight in the winter. For this reason it digs resident burrows on a steep land with a slope between 30 and 60° covered with dense shrub, canopy cover, near to water sources and that receives direct sunlight [13]. The study done by [14] also found the similar behavior of Chinese pangolin in Nepal.

4. Habitat

The potential habitat of Chinese pangolin ranges from tropical plain landscape to hilly regions. In some cases, the habitat is also found in high mountains like the northern parts of Nepal. This species is found in a wide range of habitats that include primary and secondary tropical forests, lime stone forests, bamboo forests, grasslands, agricultural fields, limestone, hill forests, broad-leaf forests in lower altitude and coniferous forests in higher attitudinal zones [6, 14–17]. The habitat is generally found in red soil at open forest with less coverage in south, south-east, and south-west facing slopes [18]. Both protected areas and areas outside of protected areas are the potential habitats of Chinese pangolin. However, most of the potential habitats are found outside of protected areas as the species is present mostly in cultivable land (**Figure 2**). Agricultural land, human settlement areas, and parts of private forests are potential habitats of Chinese pangolin. Unlike arboreal Sunda pangolin, Chinese pangolin is entirely terrestrial and digs own burrows [19, 20].

The suitable habitat of Chinese pangolin ranges within the section of mid-hills [11, 14, 16, 17]. However, some studies have found that the burrows of the species in lower altitude of below 100 m in Taiwan and exceptionally high altitude of 3000 m from sea level in eastern Nepal [2]. These studies depict that Chinese pangolin are distributed in wide range of habitats. Chinese pangolin tolerates some kind of human disturbances, therefore it prefers burrowing in the habitats near to human settlement and agricultural lands [14, 19, 21, 22]. The study conducted by [14] recorded the presence of burrows in grasslands and croplands near human settlements. Similarly, the species was sighted in evergreen coniferous forest such as pine forest, broad leaf forests such as teak *(Tectona grandis)* forest, bamboo forest, and scrubland [13, 21, 23].

Figure 2.
Burrow of Chinese pangolin found near the settlement area. Source: Author.

5. Population

Because of the species's nocturnal, elusive, secretive, and solitary nature, the direct observation is very rare unlike other mammals due to which estimating the exact number of Chinese pangolin population is extremely tough [2, 6]. Across the scales (global through local) the information on the population of Chinese pangolin is very rare along with the studies on habitat and abundance of species. This species gets little care in studies and conservation compared to other big mammals, therefore the immediate information on the population status is very low.

Unlike Taiwan, where the Chinese pangolin population is exceptionally stable and increasing [2], other home countries have faced sharp decline in the number of Chinese pangolin because of over-exploitation of habitat resulting to extirpation of original population [23, 24]. For instance, the population of Chinese pangolin in China was dramatically declined by 94% between 1960s and 1990 due to anthropogenic factors including over exploitation [9]. China harvested around 160,000 individual of Chinese pangolin every year between 1960 and 1980 for consumptive purpose that ensued in extinction of species for commercial purpose in 1990s [25]. China estimated the population of the species to be approximately 50,000–100,000 in 2002 [26] but the National Forestry Administration (now National Forestry and Grassland Administration) estimated the population of 64,000 individuals. However, in the same year [27] resulted that the population of Chinese pangolin was declined to 25,100 and 49,450 individuals.

Taiwan experienced substantial reduction in the population of Chinese pangolin between 1950s–1970s due to over exploitation from hunting and poaching [28]. However, some parts of Taiwan has recovered the population remain stable at this time as hunting and poaching are not the major threats. Vietnam hosts a few number of Chinese pangolin population. Although, the species is seen rarely, the population of

Chinese pangolin faced steep decline in the last two decades (1990s and 2000s) [20]. In Lao PDR, the spices was observed in Nam Theun Extension PNBCA (Proposed National Biodiversity Conservation Area) and in a village in Nakai-Nam Theun NBCA during 1994–1995 [6, 29]. However, the sighting of the species is exceptionally rare at this time as the population is nearly extinct due to extreme hunting of species.

In Thailand, Myanmar, and Bhutan the information about the Chinese pangolin is little or no information is available as the population has sharply declined to extinction. Similarly, Nepal has faced decline in the population of Chinese pangolin due to hunting, poaching, habitat fragmentation, and illegal trade. The survey conducted in early 1990s estimated that the population of the species was good in number in the Royal Nagarjung Forest in Kathmandu. The national population was estimated be approximately 5000 individuals, however, current population is estimated to be declined due to several anthropogenic factors including over exploitation of resources and illegal trade [11, 16, 30–32]. The similar trend is seen in India and Bangladesh. The population of Chinese pangolin is severely decreasing in Northeast India [33] and in southern part of Bangladesh due to illegal trafficking, over exploitation, hunting, and poaching [23]. Chinese pangolin population has been expected to be disappeared to extinction in southeast Bangladesh but the existence of the species is still recorded in Chittagong Hill Tracts.

6. Distribution

Chinese pangolin is very widely distributed in the continent of Asia. The distribution mostly occur in East Asia, northern Southeast Asia, and some parts of South Asia (**Figure 3**) [2]. Broadly the population of Chinese pangolin is distributed in Nepal, India, Bhutan, Bangladesh, China, Taiwan, Vietnam, Myanmar, Lao People's Democratic Republic (Lao PDR), and Thailand.

To be confined within the nations, Chinese pangolin is widely distributed in Nepal. The species is widely distributed in the mid-hills of Nepal that ranges between 500 and 1500 m [11, 17] and is exceptionally distributed at the higher altitude of 3000 m [2]. Their existence is also recorded in the southern plane with a lower altitude. Chinese pangolin is distributed throughout the county (central, eastern, and western regions) and are highly recorded in 25 districts [34]. Some of the potential areas of Chinese pangolin distribution in Nepal are Bardia, Chitwan, Makalu Barun, Parsa, Sagarmartha and Shivapuri-Nagarjun National Parks and Annapurna, Gaurishankar and Kanchenjunga Conservation Areas [34, 35]. In addition, Nepal's geographical range distributes Chinese Pangolin in Taplejung, Ilam, Pachthar, Sinduli, Ramechhap, Kavrepalanchowk, Kathmandu, Gorkha, Bhaktapur, and Dhading [11, 16, 17, 35].

Similarly, in India Chinese pangolin is distributed in northern part of the country that is mostly bordered to Nepal. The record of the species distribution is found in West Bengal, Sikkim, Bihar, Assam, Nagaland, Tripura, Mizoram, and Meghalaya [36–38]. The distribution of this species is confined to northwest and southeastern regions of Bangladesh. The Lawachara National Park, Thakurgaon region, and Chittagong Hills have record of Chinese pangolin [23]. The study conducted by [20] has recorded the distribution of this species in northern and central Vietnam some protected areas including Cuc Phuong National Park, Khe Net Nature Reserve, Ke Go Nature Reserve, and Ba Na National Park.

Likewise, the distribution record of Chinese pangolin is very few in Lao PDR and Thailand. In some parts of north and central Lao PDR, the species is sparsely

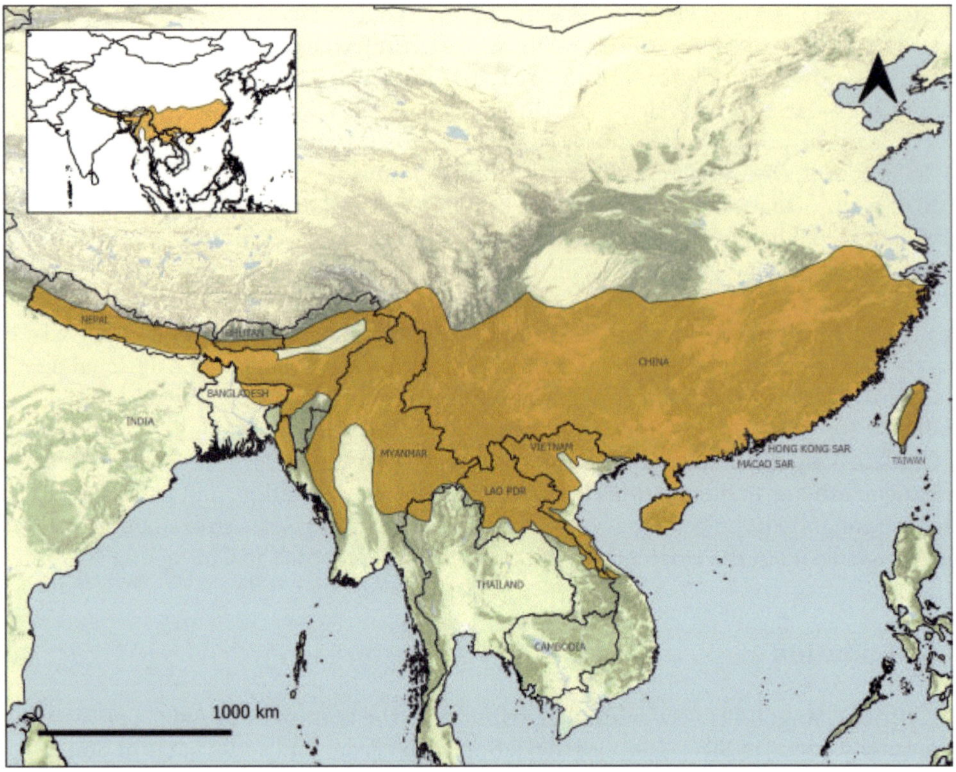

Figure 3.
Distribution of Chinese pangolin [2].

distributed [29]. A few distribution has been recorded from Changwat, Chiang Mai, and Chiang Mai in Thailand in early though mid ninety [2]. In Bhutan, Chinese pangolin is distributed mostly in southern regions. It is distributed in Samtse (southwest), Sarpang, Tsirang (central south), and Samdrup Jongkhar (southeast) in Bhutan. In Myanmar, there are records of the species distribution in northern part but the clear distribution record is not confirmed [39].

In China, the Chinese pangolin is widely distributed across southern regions. The distribution extends from Changjiang to the provinces of Yunnan, Fujian, Hunan, Jiangxi, Sichuan, Guizhou, Anhui, Jiangsu, Zhejiang, Guangdong, Huibei, Chongqing, Shanghai, and in the Autonomous Regions of Guangxi Zhuang, Tibet, and Hainan Island [2, 40]. In addition, in China the distribution of the species has been recorded in the Funiu mountain district that is located in north of Changjiang with no detailed verification and requiring further investigation [2, 6, 41]. The distribution of Chinese pangolin in Hong Kong is recorded in the periphery of lower altitudinal zone [42]. The distributional record shows that the species is found in the central and northeast New Territories and Lantau Island [6]. Unlike Hong Kong, the Chinese pangolin is distributed very widely in Taiwan. The species is found in cultivable land in lower altitude below 1000 m, and Central, Western, and Eastern Coast Mountain ranges that extends to upper limit of 2000 m [15]. The species is also distributed in the Olympics, the Taoyuan Tableland, the Tatun Volcano Group, Puli Basin, Taipei Basin, and the Pingtun Plain [2, 6, 15, 28].

7. Conservation status

With the increasing threats to the Chinese pangolin population to extinction, the global, national, and some local authorities have initiated conservation of the species. At the global scale, the IUCN Red List of Threatened Species has listed the Chinese pangolin under the Critically Endangered category [2]. Similarly, the Chinese pangolin is listed in Appendix I in the CITES. Nationally, in Nepal the species is categorized as an Endangered species and is listed as the protected species under the National Parks and Wildlife Conservation (NPWC) Act 1973 [30, 43]. Moreover, under CITES, Nepal has registered under Appendix I to restrict all kinds of wild killing and commercial use. In 2018 the Department of National Parks and Wildlife Conservation in Nepal developed a Pangolin Conservation Action Plan for Nepal (2018–2022) to promote mitigating illegal trade and the conservation of Chinese pangolin in Nepal by implementing immediate action in identifying potential habitats and the regular monitoring of populations and suitable habitats [16]. In addition, several Community Forests User Groups (CFUGs) have initiated pangolin conservation through their operational plan including the Rani CFUG in Hetauda Nepal.

Likewise, China has listed the species under Critically Endangered category on its National Red List [44]. India has listed Chinese pangolin in protected list by listing it in Schedule I of the Wildlife Protection Act 1972 and under Appendix II of CITES [33]. These conservation steps in India restricts for the wild catch and commercial trade of Chinese pangolin within and outside of country. Similarly, Taiwan, Hong Kong, Bangladesh, and Myanmar have listed Chinese pangolin under protected species list. The primary intention of these countries to protect the Chinese pangolin is to reduce wild catching, killing, poaching, and illegal trade for commercial purposes within the country and outside the national boundaries [2, 6].

8. Threats

Despite several conservation efforts implemented across scales (international, national, and some local) for a long time, Chinese pangolin population is under the severe threats that could result to the extinction of species in the near future. There exist certain natural threats, however, anthropogenic factors are primarily responsible for the steep decline in the population of Chinese pangolin and destruction of their habitats. Anthropogenic threats including illegal hunting, poaching, illegal trade, deforestation, forest fire, agriculture expansion, forest encroachment, traditional beliefs, over-grazing, and developmental activities on haphazard manner are the major reasons behind the dramatic decline of Chinese pangolin globally, nationally, and locally.

8.1 Hunting and poaching

Chinese pangolin has been poached and hunted in wild for both household and commercial purposes for several decades. At the local level the species is killed as a source of meat in some parts of the world where as the killing is done primarily for illegal trade at national and international levels [2]. The habitats that are near the human settlement and agricultural lands are highly vulnerable because hunters have easy access and they often poach and hunt Chinese pangolin in these areas (**Figure 4**)

Figure 4.
The scales of locally killed Chinese pangolin. Source: Badri Binod Dahal.

[17, 27]. Evidences and studies claim that Chinese pangolin has been historically exploited for medicinal, cultural, spiritual, and nutritional purposes [2, 23, 24, 45]. Local people consume on Chinese pangolin with a traditional belief of medicinal values that heals several kinds of diseases such as skin disease, back pain, gastrointestinal disease, pain killer during pregnancy, and cardiac problem [46].

The meat of Chinese pangolin is consumed as a source of protein. Similarly, the study done by [46, 47] have recorded a significant number of respondents were aware about the protection status of Chinese pangolin but they were had also eaten the meat of the species. Likewise, in the eastern Nepal the Limbu community have negative attitude towards Chinese pangolin and perceive it as a symbol of bad luck and kill it wherever they see the species [46, 48]. Some tribal communities in northeastern India consume the meat of at the community level and they use scales for several medicinal purposes [33, 45]. The scales of the species was also used to make decorative items, rings, musical instruments, and bags at the community level that has enhanced the threats to the Chinese pangolin population. In China and Vietnam, the scales of Chinese pangolin is considered as a source of traditional medicine for treating skin diseases, treating cancer, blood circulation, and stimulating milk secretion in lactating women [2, 45].

8.2 Illegal trade

Illegal trade of a live and dead Chinese pangolin along with its body parts including scales and meat is the primary threat to the remaining population of this species [46]. Usually the illegal trade of Chinese pangolin happens at the international scales because of an excessive demand of the species in China. The local hunters work actively and with full effort in catching the Chinese pangolin alive or killing to supply the demand across the scales that has very sharply declined the number of the species in the past few decades [6, 32, 47]. There is a long chain of hunters and traders in most of the countries that host Chinese pangolin and they are connected internationally to promote illegal hunting, trafficking and trading of the species that has led the species in the verge of extinction in the contemporary world (**Figure 5**) [32, 49].

In the last 21 years, the population of Chinese pangolin has been estimated to be declined by more than 90% because of escalated anthropogenic activities, illegal trade in particular. Pangolin is the most widely illegally traded species in the south-eastern Asia because of its demand in China for multiple purposes. Historically the illegal trade of the species has been recorded by CITES. A large population of Chinese pangolin declined in mid-1990s due to unrecorded internationalized illegal trade of a live species that had occurred between Southeast Asia and China along with other species [2, 20, 50]. Most of the illegally traded Chinese pangolin to China originates from Myanmar and Nepal [32, 51]. During the decades of 1980s and 1990s, CITES reported that international trade of more than 50,000 (estimated) Chinese pangolin were internationally traded [6].

Although CITES banned in illegal trade of Chinese pangolin in 2000 by introducing zero export of this species following the severe decline in the Chinese pangolin number in wild due to escalated illegal trade, illegal trade and trafficking of the species has continued post-2000 that is governed and driven by market demand of live individuals in China from Southeast and South Asia [49, 51]. In the time difference of 3 years (between 2000 and 2013), the seizure records of Chinese pangolin show that more than 50,000 Chinese pangolin were internationally trafficked [49, 52]. The high demand and conspicuously consumption of Chinese pangolin as a luxury wild meat

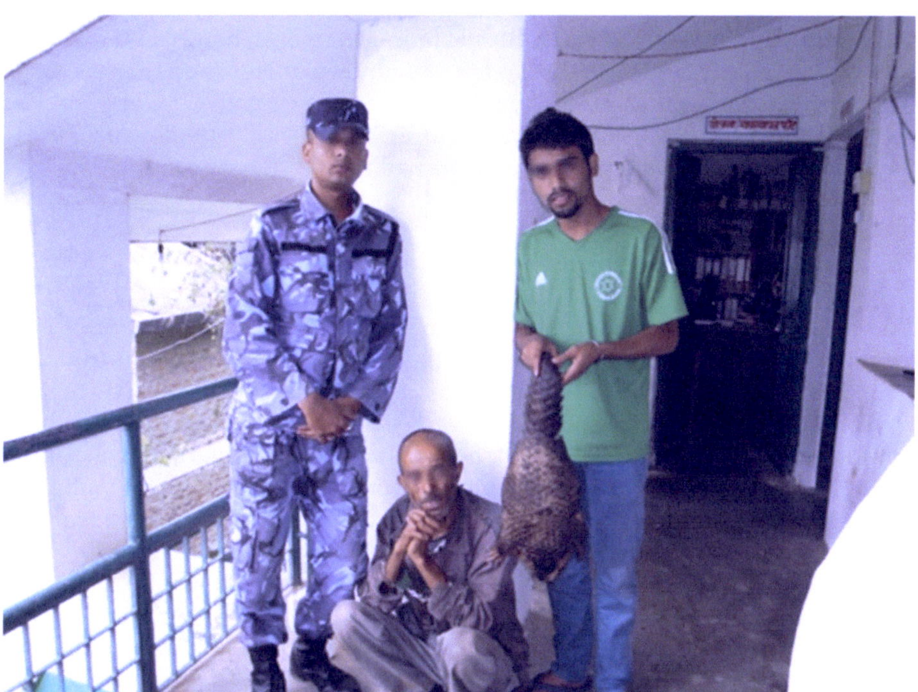

Figure 5.
Local hunter with dead pangolin. Source: Author.

dish among opulent consumers in China and Vietnam has driven illegal international trade of the species in high prices [2, 49].

8.3 Over exploitation and habitat fragmentation

Apart from illegal trade, hunting and poaching, anthropogenic factors are highly responsible for over exploitation of resources and habitat fragmentation that has resulted in the extirpation of Chinese pangolin from different regions including Bangladesh and some provinces of China [6, 23]. Most of the countries in Asia that are also the home countries to Chinese pangolin are currently on a developing phase that results in over exploitation of resources such as forests, wetlands, agricultural lands, grasslands, and shrub-lands for constructions and other developmental activities. These commercial developmental activities resulting to over exploitation of resources either destroy the potential habitats of the species entirely or fragment into parts that will directly result in the decrement of Chinese pangolin population in the wild [6, 53].

In the countries like Nepal, India, and Bangladesh, the rural areas are in the phase of rapid conversion to semi-urban or urban areas by destroying forests, agricultural lands, and barren lands that are primary habitat of Chinese pangolin [11, 16, 22, 23, 53]. In addition, the human settlement has expanded by encroaching nearby forests and agricultural lands. These activities at the household and developmental scales have exploited resources beyond limits in one hand and increased severely threats to the habitat on the other hand that has resulted in reduced number of Chinese pangolin. Some examples of developmental activities such as large scale road network constructions, industrial areas construction, infrastructure construction, mining etc. have fragmented or entirely destroyed the potential habitats of Chinese pangolin. The study done by [17, 53] revealed that the Balthali village is currently suffering from speed conversion of forest and agricultural lands into resorts, hotels, and road constructions that have fragmented the habitat of Chinese pangolin and this has resulted in decline of population. In addition to these threats, the excessive use of pesticides, insecticides and other chemicals in the agriculture lands with pangolin burrows is a severe threat to both Chinese pangolin and its habitat [6, 53].

9. Recommendations in Chinese pangolin studies

There are considerable studies done on Chinese pangolin in different parts of Asia where this species is found. For instance, the studies in distribution based on direct and indirect signs of pangolin, conservation, threats, spaces abundance, illegal trade based on seizure records, and habitat preference have been conducted. In addition, the conservation efforts to conserve the species have been implemented at international, national, and some local levels. Despite these studies and efforts, the information on multiple dimensions of Chinese pangolin still lacks in research arena and the threats to conservation of the species is still up-scaling. This section in this chapter tries to provide certain recommendations in Chinese pangolin studies and conservation.

i. Most of the researches on Chinese pangolin has concluded that the studies on the population is lacking. It is very important to have a scientific data on the population of the species to support all kind of research and focus conservation efforts. Therefore, a comprehensive study on the population of Chinese pangolin is of a dire need.

ii. There is a lack of behavioral studies of Chinese pangolin because of its unique nature. However, further researches should be able to develop a new techniques and methods to comprehend the ecological and other behaviors of Chinese pangolin. This will also help to understand the distribution, habitat preference, and abundance of the species based on the direct signs as most of the current studies are based on indirect signs.

iii. Although the issue of climate change in all sectors are alarming, there is no single study on the impacts of climate change on Chinese pangolin population and its habitats. It is a crucial topic that should be studied immediately in the contemporary world that is apparently facing multiple impacts of climate change in environmental sectors and biodiversity including small mammals like Chinese pangolin and its habitats.

iv. As the conservation of wildlife species have become a clear political agenda, the political dimensions of Chinese pangolin conservation is very crucial subject of research. In addition, the Chinese pangolin might be threatened at local level as the local/national politics might influence local level poaching and hunting and national/international political sphere might influence the global market of illegal trade and trafficking.

v. The clear records on the current illegal trading of Chinese pangolin and the estimated projection of future illegal trade is lacking. Future researches should focus on comprehensive research in quantifying hunting pressure and use of the species at local, national, and international scales.

vi. The conservation efforts should be designed and implemented for an impact results in conserving the species and its habitat rather than just being limited within the texts. For example, Nepal developed a Pangolin Conservation Action Plan (2018–2022) to identify potential habitats, regularly monitor population and habitats, mitigate illegal hunting and trading. However, the comprehensive review of the action plan to understand the impacts of the plan over its planned objectives has not been publicized yet. In addition, the conservation output from implementing expensive conservation action plan has not been produced yet. Therefore, the conservation efforts should be designed and implemented for an effective outputs.

Community based/initiated conservation programs should be designed and implemented rather than imposing conservation programs designed by authorities. This will encourage community people and enhance their understanding capacity regarding Chinese pangolin conservation. Handing over of the power and authority in developing, implementing, and monitoring local action plans will enhance sustainable conservation of the species.

Conflict of interest

The author declare that there is no any conflict of interest.

Author details

Suman Acharya[1,2,3]

1 Anthropology and Environmental Policy, Department of Anthropology, University of Maine, Orono, ME, USA

2 Climate Change Institute, University of Maine, Orono, ME, USA

3 Himalayan Conservation and Research Institute, Nepal

*Address all correspondence to: acharya.suman1994@gmail.com

References

[1] Dickman CR, Richer RA. In: Macdonald DW, editor. Pangolins. The Encyclopedia of Mammals. New York: Facts on File; 2001. pp. 800-801

[2] Challender D, Wu S, Kaspal P, Khatiwada A, Ghose A, Ching-Min Su N, et al. Manis pentadactyla. The IUCN Red List of Threatened Species 20192019. DOI: 10.2305/IUCN.UK.20193.RLTS. T12764A123585318.en [Accessed: August 8, 2023]

[3] CITES. The CITES appendices, Appendices I, II & III. 2019. Available from: https://cites.org/sites/default/files/eng/app/2017/E-Appendices-2017-10-04.pdf [Accessed: August 12, 2023]

[4] Gaubert P, Antunes A, Meng H, Miao L, Peigné S, Justy F, et al. The complete phylogeny of pangolins: Scaling up resources for the molecular tracing of the most trafficked mammals on earth. Journal of Heredity. 2018;**109**(4):347-359. DOI: 10.1093/jhered/esx097

[5] Gotch AF. Mammals, their Latin Names Explained: A Guide to Animal Classification. Poole: Blandford Press; 1979

[6] Wu S, Sun NCM, Zhang F, Yu Y, Ades G, Suwal TL, et al. Chinese pangolin Manis pentadactyla (Linnaeus, 1758). In: Pangolins. London: Academic Press; 2020. pp. 49-70. DOI: 10.1016/B978-0-12-815507-3.00004-6

[7] Pocock RI. The external characters of: The pangolins (Manidæ). In: Proceedings of the Zoological Society of London. Oxford, UK: Blackwell Publishing Ltd; 1924. pp. 707-723. DOI: 10.1111/j.1096-3642.1924.tb03310.x

[8] Heath ME, Vanderlip SL. Biology, husbandry, and veterinary care of captive Chinese pangolins (*Manis pentadactyla*). Zoo Biology. 1988;**7**(4):293-312. DOI: 10.1002/zoo.1430070402

[9] Wu SB, Liu NF, Zhang YM, Ma GZ. Physical measurement and comparison for two species of pangolin. Acta Theriologica Sinica. 2004;**24**(4):361-364

[10] Heath ME. Manis pentadactyla. Mammalian Species. 1992;**414**:1-6

[11] Suwal TL, Thapa A, Gurung S, Aryal PC, Basnet H, Basnet K, et al. Predicting the potential distribution and habitat variables associated with pangolins in Nepal. Global Ecology and Conservation. 2020;**23**:e01049. DOI: 10.1016/j.gecco.2020.e01049

[12] Lin MF, Chang CY, Yang CW, Dierenfeld ES. Aspects of digestive anatomy, feed intake and digestion in the Chinese pangolin (*Manis pentadactyla*) at Taipei zoo. Zoo Biology. 2015;**34**(3):262-270. DOI: 10.1002/zoo.21212

[13] Wu SB, Liu NF, Ma GZ, Xu ZR, Chen H. Habitat selection by Chinese pangolin (Manis pentadactyla) in winter in Dawuling Natural Reserve. Mammalia. 2003;**67**(4):493-502. DOI: 10.1515/mamm-2003-0403

[14] Gurung JB. A pangolin survey in Royal Nagarjung forest in Kathmandu, Nepal. Tigerpaper. 1996;**23**(2):29-32

[15] Chao J. Studies on the Conservation of the Formosan Pangolin (*Manis pentadactyla*). I. General Biology and Current Status. Executive Yuan, Taiwan: Division of Forest Biology, Taiwan Forestry Research Institute; 1989

[16] Sharma HP, Rimal B, Zhang M, Sharma S, Poudyal LP, Maharjan S, et al.

Potential distribution of the critically endangered Chinese Pangolin (Manis pentadactyla) in different land covers of Nepal: Implications for conservation. Sustainability. 2020;**12**(3):1282. DOI: 10.3390/su12031282

[17] Acharya S, Sharma HP, Bhattarai R, Poudyal B, Sharma S, Upadhaya S. Distribution and habitat preferences of the Chinese Pangolin Manis pentadactyla (Mammalia: Manidae) in the mid-hills of Nepal. Journal of Threatened Taxa. 2021;**13**(8):18959-18966. DOI: 10.11609/jott.3952.13.8.18959-18966

[18] Shrestha B. Distribution and Diversity of Mammals in Shivapuri National Parks. Nepal: Tribhuvan University; 2005

[19] Wu S. Observation on food habits and foraging behavior of Chinese pangolin (*Manis pentadactyla*). Chinese Journal of Applied and Environmental Biology. 2005;**11**(3):337

[20] Newton P, Van Thai N, Roberton S, Bell D. Pangolins in peril: Using local hunters' knowledge to conserve elusive species in Vietnam. Endangered Species Research. 2008;**6**(1):41-53. DOI: 10.3354/esr00127

[21] Kaspal P. Status, Distribution, Habitat Utilization and Conservation of Chinese Pangolin in the Community Forests of Suryabinayak Range Post. Nepal: Khowpa College; 2008

[22] Acharya S. Assessment of Conservation Threats and Habitat Management of Chinese Pangolin (*Manis pentadactyla*) in Balthali VDC of Kavreplanchowk Nepal. Pokhara, Nepal: Institute of Forestry, Tribhuvan University; 2016

[23] Trageser SJ, Ghose A, Faisal M, Mro P, Mro P, Rahman SC. Pangolin distribution and conservation status in Bangladesh. PLoS One. 2017;**12**(4):e0175450. DOI: 10.1371/journal.pone.0175450

[24] Nash HC, Wong MH, Turvey ST. Using local ecological knowledge to determine status and threats of the critically endangered Chinese pangolin (Manis pentadactyla) in Hainan, China. Biological Conservation. 2016;**196**:189-195. DOI: 10.1016/j.biocon.2016.02.025

[25] Zhang L, Hua N, Sun S. Wildlife trade, consumption and conservation awareness in Southwest China. Biodiversity and Conservation. 2008;**17**:1493-1516. DOI: 10.1007/s10531-008-9358-8

[26] Wu SBMG, Ma GZ, Tang M, Chen H, Liu N. The status and conservation strategy of pangolin resource in China. Journal of Natural Resources. 2002;**17**(2):174-180

[27] Zhang L, Li Q, Sun G, Luo S. Population status and conservation of pangolins in China. Bulletin of Biology. 2010;**45**:1-4

[28] Chao JT, Tsao EH, Traylor-Holzer K, Reed D, Leus K. Formosan Pangolin Population and Habitat Viability Assessment: Final Report. Apple Valley, MN, USA: IUCN/SSC Conservation Breeding Specialist Group; 2005

[29] Duckworth JW, Salter RE, Khounbline K. Wildlife in Lao PDR: 1999 Status Report. Vientiane, Laos: IUCN; 1999

[30] Jnawali SR, Baral HS, Lee S, Subedi N, Acharya KP, Upadhyay GP, et al. The Status of Nepal's Mammals: The National Red List Series. Kathmandu, Nepal: Depart of National Parks and Wildlife Conservation; 2011

[31] Sharma S, Acharya S. Human-rhesus macaque conflict at pumdivumdi/

tallokodi, Pokhara, West Nepal. Banko Janakari. 2017;**27**(2):46-50

[32] Ghimire P, Raut N, Khanal P, Acharya S, Upadhaya S. Species in peril: Assessing the status of the trade in pangolins in Nepal. Journal of Threatened Taxa. 2020;**12**(8):15776-15783. DOI: 10.11609/jott.5698.12.8.15776-15783

[33] Mohapatra RK, Panda S, Acharjyo LN, Nair MV, Challender DW. A note on the illegal trade and use of pangolin body parts in India. Traffic Bulletin. 2015;**27**(1):33-40

[34] DNPWC; DoF. Pangolin Conservation Action Plan for Nepal (2018-2022). Kathmandu, Nepal: Department of National Parks and Wildlife Conservation and Department of Forests; 2018

[35] Shrestha TK. Wildlife of Nepal: A study of Renewable Resources of Nepal, Himalayas. Kathmandu: Curriculum Development Centre, Tribhuvan University; 1981

[36] Misra M, Hanfee N. Pangolin distribution and trade in East and Northeast India. Traffic Dispatches. 2000;**14**:4-5

[37] Srinivasulu C, Srinivasulu B. South Asian Mammals: Their Diversity, Distribution, and Status. New York: Springer; 2012. DOI: 10.1007/978-1-4614-3449-8

[38] Maurya KK, Shafi S, Gupta M. Chinese Pangolin: Sighting of Chinese Pangolin (*Manis pentadactyla*) in Valmiki Tiger Reserve, Bihar, India. Small Mammal Mail. 2018;**33**(1):15-18

[39] Corbet GB, Hill JE. The Mammals of the Indomalayan Region: A Systematic Review. Vol. 488. Oxford: Oxford University Press; 1992

[40] Allen GM, Coolidge HJ. Mammal and bird collections of the Asiatic primate expedition. Mammals Bulletin Mus. Comp. Zool. Harv. 1940;**87**:131-166

[41] Zhai W. Rare and Endangered Animals in Henan Province. Zhengzhou: Henan Science and Technology Press; 2000

[42] Reels G. Distribution of Large Mammals in Hong Kong: A Summary of Sightings Made since 1992. Hong Kong: Hong Kong University; 1996

[43] DNPWC. 2022. Department of National Parks and Wildlife Conservation. Available from: https://dnpwc.gov.np/en/speciesdetail/55/#:_x0005_:text=There%20is%20very%20little%20information,observed%20to%20 be%20in%20decline [Accessed: August 12, 2023]

[44] Jiang Z, Jiang J, Wang Y, Zhang E, Zhang Y, Li L, et al. Red list of China's vertebrates. Biodiversity Science. 2016;**24**(5):500

[45] D'Cruze N, Singh B, Mookerjee A, Harrington LA, Macdonald DW. A socio-economic survey of pangolin hunting in Assam, Northeast India. Nature Conservation. 2018;**30**:83-105. DOI: 10.3897/natureconservation.30.27379

[46] Katuwal HB, Neupane KR, Adhikari D, Thapa S. Pangolins Trade, Ethnic Importance and its Conservation in Eastern Nepal. Kathmandu, Nepal: Small Mammals Conservation and Research Foundation and WWF-Nepal; 2013

[47] Katuwal HB, Neupane KR, Adhikari D, Sharma M, Thapa S. Pangolins in eastern Nepal: Trade and ethno-medicinal importance. Journal of Threatened Taxa. 2015;**7**(9):7563-7567. DOI: 10.11609/JoTT.o4202.7563-7

[48] Acharya S. Pains of the Pangolin. The Kathmandu Post. 2015. Available from: htps:// kathmandupost.com/ opinion/2015/02/22/pains-of-the-pangolin/ [Accessed: August 8, 2023]

[49] Challender DW, Heinrich S, Shepherd CR, Katsis LK. International trade and trafficking in pangolins, 1900-2019. In: Pangolins. London: Academic Press; 2020. pp. 259-276. DOI: 10.1016/ B978-0-12-815507-3.00016-2

[50] Wu SB, Ma GZ. The status and conservation of pangolins in China. TRAFFIC East Asia Newsletter. 2007;**4**:1-5

[51] Challender D, Waterman C. Implementation of CITES Decisions 17.239 B and 17.240 on Pangolins (Manis spp.). Prepared by IUCN for the CITES Secretariat. SC69 Doc, 57, 19-21. 2017

[52] Challender DW, Harrop SR, MacMillan DC. Understanding markets to conserve trade-threatened species in CITES. Biological Conservation. 2015;**187**:249-259. DOI: 10.1016/j. biocon.2015.04.015

[53] Acharya S, Rayamajhi S, Sharma S, Upadhaya S, Joshi S, Lamichhane S. Anthropogenic threats to survival of the critically endangered Chinese pangolins (*Manis pentadactyla*) and their habitat in Kavrepalanchowk Nepal. Journal of Biodiversity & Endangered Species. 2018;**6**(218):2. DOI: 10.4172/2332-2543.1000218

Section 2

Conservation and Challenges of Endangered Plant Species

Chapter 4

Research Advances and Perspectives of Conservation Genomics of Endangered Plants

Qing Ma, Gang Wu, Wenjie Li, Seyit Yuzuak, Fachun Guan and Yin Lu

Abstract

Understanding in the evolutionary processes, endangered mechanisms, and adaptive evolution history are key scientific issues in conservation biology. During the past decades, advances in high-throughput sequencing and multi-disciplinary crossover have triggered the development of conservation genomics, which refers to the use of new genomic technologies and genomic information in solving the existing problems in conservation biology. Conservation genomics mainly focuses on the endangered mechanism and conservation strategies aiming at protection of survivability and diversity of endangered species. Application of conservation genomics into the study of endanger plant species has provided innovated protection concept for biologists and promoted the development of population-based conservation strategies. This chapter summarizes the studies of population genomics for agronomically and commercially important plants threatened and endangered, discusses the advantages of conservation genomics for the analysis of genetic diversity, inferences about the history of population dynamics, evaluation of natural forces on wild plant populations, and the establishment of effective conservation strategies. This chapter also presents the development trends in genomics for the conservation of endangered plant species.

Keywords: endangered species, genomics, conservation strategy, genetic diversity, conservation biology

1. Introduction

In modern times, frequent and intense human activities and habitat degradation have caused major and serious threats on the plant's biodiversity [1, 2]. Global energy consumption is becoming increasingly severe, and the global climate and environment is undergoing drastic changes. The size of existing populations is rapidly shrinking and many species are on the brink of extinction. The resulting population bottleneck further reduced the adaptive potential and functional diversity of species, and poses dangers and difficulties in the recovery of populations under natural conditions. Many wild animals and plants are facing an unprecedented survival crisis [3–7].

In response to this crisis, conservation biology has emerged. After years of development in conservation biology, preservation of biodiversity in the context of ecosystem has arrived at a consensus, and so the mechanism and priciple factors responsible for biodiversty loss has been a hot topic in the field of conservation biology. The requirement of better evaluation of species endangerment degree, more accurate estimation of endangered species habitats, and more suitable conservation strategies has led to the development and application of multiple related technologies, such as geographic positioning system (GPS) technology, mathematical methods, and genomics study [8]. In recent decades, the rapid development of high-throughput technology has greatly increased the availability of genomic data and effectively solved the management problems of many endangered species and related key protection regions.

2. An overview of conservation genomics

2.1 Main features

As a branch of conservation biology, conservation genomics focuses on the study of endangered mechanisms and conservation strategies for endangered species, that aims to protect species survival ability and reduce the risk of extinction. This field is an integration of theoretical ideas of genetics and analytical methods of genomics.

Traditional conservation genetics research often relies on only a few molecular markers such as isozyme, microsatellites, or mitochondrial and chloroplast genomes. Since most of these molecular markers are anchored in a few neutral regions of the genome, there are certain limitations in tackling with questions in species conservation [9, 10]. Compared with conservation genetics, conservation genomics is more direct and effective in addressing some important issues in conservation biology.

For example, when estimating effective population size (Ne), genome data can provide a large number of genetic markers, which can be used to reconstruct pedigree and extract haplotype information, with which Ne values can be monitored, population migration direction predicted, and migrating individuals can be identified. When inferring the adverse effects of inbreeding depression, whole genome information can directly locate the key sites, accurately predict the population's ability to eliminate harmful mutations, and screen the initial founders of captive populations based on the genotype of individual inbreeding sites.

Furthermore, when estimating the interference of climate change and human activities on wild populations, the large amount of genetic variation information obtained through genomic sequencing can help accurately evaluate the responsive ability of different individuals and help to find individuals suitable for ex-situ conservation [11]. In fact, these issues have long been well-recognized in conservation biology research, but due to the lack of key genomic information, there has been no further development until the advent of next-generation sequencing, which successfully opened up new research perspectives.

2.2 Significance in conservation biology

In the past 200 million years, an average of 900,000 species of animals have become extinct per million years, with a "background rate" of approximately 90 species per century, including approximately four species of higher plants [12]. In history, the five mass extinctions that have occurred to date have generally occurred as a result of significant geological events or rapid environmental changes. Studies on general biodiversity loss indicate that we are now in the sixth mass extinction [13]. It is conservatively estimated that the extinction rate of species in the past century has been 22 times faster than historical benchmarks [14].

In 2022, 11,538 plant species were included in the red list updated by the International Union for Conservation of Nature and Natural Resources (IUCN), of which about 5336 (46.2%) were critically endangered, 10,202 (88.4) were endangered, 9376 (81.2%) were vulnerable, 134 (1.2%) were conservation dependent, and 3761 (32.6%) were near threatened [15]. China is one of the countries with the richest biodiversity in the world, with about 30,000 species of higher plants, more than 5% of which are endemic, and ranks third in the world. According to statistics reported so far, there are 4000 to 5000 plant species in China that are endangered or threatened, accounting for 15% -20% of all plant species. Nearly 100 species are facing extinction. Therefore, the application of modern multidisciplinary research methods is of great importance in order to protect and strengthen plant germplasm resources and maintain plant diversity.

Conservation genomics not only provides more comprehensive, accurate and reliable research results on species classification, genetic diversity, population genetic structure and so on, but also provide insights into the historical processes of species origin, differentiation, population size evolution, and the molecular mechanism of population local adaptation, inbreeding status and genetic basis of inbreeding depression [11, 16, 17]. Various models based on conservation genomics have been developed to directly identify biodiversity hotspots to provide priority protection.

In addition, new classes of markers obtained from whole genome sequence can be used to screen for functional genes and key adaptive sites contributing to important ecological adaptations, such as the stresses of climate change, resistance to herbivory, and disease, making it possible to predict species' response to the environment in the past and future [18].

Beginning from the studies of traditional genetic diversity, genetic structure, and population dynamics, conservation genomics allows us to delve into the reconstruction of evolutionary history and species adaptive evolution and explore the process, causes, and evolutionary potential of endangerment. Application of genomics study in conservation biology can guide practical management actions of endangered species from both the spatial and temporal perspectives.

2.3 Research techniques and strategies

2.3.1 Development of genome sequencing technology

Genome sequencing technology has gone through a development process from first generation to third generation sequencing. The first generation sequencing is also known as Sanger sequencing. Its core principle is to image different lengths of DNA fragments containing isotope markers through gel electrophoresis during DNA synthesis and identify the type of DNA base at each position [19]. The first generation sequencing has the advantages of long read length and high accuracy, but its low throughput, long sequencing time and high cost made it unsuitable for large-scale genome sequencing.

The second generation sequencing or next generation high-throughput sequencing determines the DNA sequence by capturing the special markers (usually fluorescent molecular markers) carried by the newly added bases during DNA replication [20]. It has advantages such as high throughput, fast speed, and low cost, but there are also shortcomings such as limited read length and assembly fragmentation.

The third generation sequencing technology, also known as single molecule real-time sequencing technology can obtain DNA base information in real-time through captured optical or electrical signals [21, 22]. Its biggest feature is single molecule sequencing, which does not require PCR amplification during the sequencing process. Together with its advantages of long read length (10–150 kb), fast speed, and no GC preference, the third generation sequencing greatly improves the integrity of genome assembly [23]. However, a disadvantage of this technology is the relatively high error rate of single base sequencing, which can reach 15%. It may require the use of second-generation sequencing data to correct the sequenced bases.

2.3.2 Simplifying genome sequencing

At present, two main categories of research techniques are widely used in conservation genomics: simplified genome sequencing and whole genome sequencing [16].

Simplifying genome sequencing is one of the commonly used techniques. It means sequencing of partial genomes, which greatly reduces the complexity of the genome, thereby lowering the cost and computational burden of sequencing. Simplifying genome sequencing has many advantages over whole genome sequencing, such as cost-effective, good stability, shorter experimental time, simpler library construction program, gain of a larger number of SNPs (single nucleotide polymorphisms), and independence of the reference genome. Hence, this technology is widely used in the conservation genomics studies of endangered animals and plants [16, 24–26]. Simplified genome sequencing can be divided into restriction site associated DNA sequence (RAD-seq) [27], RNA transcriptome sequencing (RNA-seq) [28], and whole exome sequencing (WES) [29]. The commonality of these three methods is that they normally only evaluate a small portion of the genome. However, due to the incomplete genome coverage and missing data, simplifying genome data poses challenges for subsequent population genetic analysis, such as in the inference of population phylogenetic relationship. Firstly, when there are polymorphisms or sequencing errors, it is difficult to conduct precise cluster analysis of the same restrictive loci. Secondly, it is still quite complicated to assemble each gene cluster into unique loci and ultimately construct phylogenetic relationships. Finally, the scale of genetic variation information obtained from RAD seq and the availability of RAD-seq data in phylogeny inference are influenced by various factors such as the restriction enzymes used, the size of selected fragments, and the sequencing coverage of different samples [27, 30, 31]. By contrast, whole genome re-sequencing method relying on the reference genome can significantly improve the quantity and quality of detected genetic markers [27, 32].

2.3.3 Whole genome sequencing

Whole genome sequencing can be divided into two categories: De novo whole genome sequencing and whole genome re-sequencing. De novo sequencing refers to the first assembly of a new genome sequence. The difficulty and quality of genome assembly depend on genome size, complexity, computational conditions, and bioinformatics techniques. Currently, De novo whole genome sequencing mainly utilizes third-generation sequencing techniques, including single molecule nanopore DNA sequencing from Oxford Nanopore Technologies (ONT), single molecule real-time sequencing (SMRT) from Pacific Biosciences (PacBio), and true single molecular sequencing (tSMS) from Helicos Biosciences.

In comparison, whole genome re-sequencing aims at comparison of genomic variation among different individuals and populations based on genomic sequencing of different individuals of species with known genome sequences. It mainly utilizes second-generation sequencing techniques, such as 454 GS FLX Titanium Platform supplied by Roche, HiSeq 2000 Platform supplied by Illumina, and the ABI SOLiD Platform to obtain a large number of short reads. After comparing with the reference genome, population level single nucleotide polymorphism (SNP) data can be obtained and used for population genetic analyses.

Whole genome re-sequencing requires high-quality reference genomes for read length alignment and mutation detection. The lack of high-quality reference genomes is the main obstacle for the use of whole genome re-sequencing technology in conservation biology. Despite the rapid development of sequencing technology, whole genome data is still unavailable in many endangered species. A statistical analysis has shown that among all the plant species with whole genome data, only 3.25% were threatened species included in IUCN red list and only 5.34% were included in the List of Wild Plants Under State Protection in China [33]. Nevertheless, with the increasing awareness of conservation and the launch of some important projects, it is expected that more and more endangered plant genomes will be analyzed in the future. For example, the Earth BioGenome Project proposed to give priority to sequencing the genomes of more than 23,000 endangered species included in the IUCN Red List [34]. The implementation of this project will provide assistance for the conservation genomics research of endangered species.

3. Application of conservation genomics

3.1 Determination of conservation units

In conservation biology, 'species' is commonly used to express the concept of conservation units, including taxonomic levels below species, such as subspecies and populations. It is very important for biodiversity conservation to determine population units that meet the requirements of taxonomy for standardized management. The successful implementation of the conservation plan largely depends on the correct identification of the taxonomic status of conservation targets [35]. In conservation biology, researchers tend to define species based on phylogenetic analysis. However, some different subspecies may be mistakenly treated as a different species. Other widely accepted definition methods require calculation of genetic distance or prove for reproductive isolation. In certain cases, threatened species may be described using minor morphological or distributional features, which brings controversial opinions on the taxonomic boundaries and the necessity of conservation.

Whole genome records the entire history of evolutionary process of each species. By comparing the genomic data rather than a few genes like traditional genetic analytical methods, more robust phylogenetic relationships can be constructed, providing new solutions for the identification of closely-related species and the discovery of hidden species [16].

For example, the rare and endangered plant *Buddleja alternifoli* (Scrophulariaceae) in Inner Mongolia is mainly distributed in the three major regions of the Himalayas, Hengduan Mountains, and the Loess Plateau. Significant morphological differences in between the populations in the Loess Plateau region and the other two regions were detected, while the populations in the Himalayas and Hengduan Mountains have no evident differences in morphology, thereby raising the doubt whether the populations distributed in the three regions belong to one species. Ma et al. first assembled the high-quality genome of *Buddleja alternifoli* and obtained sample re-sequencing data from 48 populations distributed in three regions. They found that *Buddleja alternifoli* formed three independent and distinct branches consistent with geographical distribution. The population differentiation coefficient FST was greater than 0.5. They speculated that *Buddleja alternifoli* in these three regions should be defined as three different species. Given the fact that *Buddleja alternifoli* is already an endangered species, results from genomic analysis implicated that each newly defined species has fewer numbers and narrower distribution range than previous prediction. The actual endangered degree of the species may be higher. When implementing protection management, these three possible species should be managed separately [36]. Similarly, although some endangered plants cannot be determined from their morphology whether they belong to different species, genetic differentiation is already very large and can actually be located as different species.

On the contrary, mis-identification of taxonomic status may cause unnecessary conservation actions and costs. *Banksia vincentia* was originally identified as a member of a species complex constitutes by *Backobourkia collina*, *B. cunninghamii*, *B. neoanglica*. It was treated as a critically endangered species in New South Wales, Australia. However, nuclear genomic data did not support '*B. vincentia*' as a distinct species and showed that it is nested within *B. neoanglica*. The value of conservation of '*B. vincentia*' needs to be further evaluated [37].

Differentiation of genetic diversity is often existed between different populations of the same species. One of the goals of conservation biology is to protect the genetic diversity of vulnerable species as much as possible. The most important step in population management is to determine and delineate the boundaries of conservation units (CUs) within species, such as evolutionarily significant units (ESUs). If a population basically has reproductive isolation with other populations of the same species, and represents an important evolutionary component of the species, then the population can be regarded as an ESU [38]. The significant units of evolution represent the vast majority of genetic diversity between populations within a species [39, 40]. In addition to ESUs, there are also management units (MUs) and adaptive units (AUs). By dividing these population units within a species, each population can receive targeted and efficient supervision, including reasonable planning of harvest yield to avoid excessive harvest, introduction of new individuals to the population to avoid mixing populations with different adaptations, and prioritize protection for certain population units to save budgets.

Within the genomic framework, it was suggested that genetic structure analysis of all possible loci to should be used to identify ESUs, neutral loci should be used to identify MUs, and adaptive loci should be used to identify AUs [39]. At present, using whole genome sequencing combined with re-sequencing data for population genetic structure analysis is a routine in population genomics [36, 41, 42]. The results can serve as a basis for ESUs classification of endangered plants. Using whole genome de novo sequencing, chromosome level assembly, and population re-sequencing of the endangered species *Tetracentron sinensis*, biologists found that 55 individuals of the species were distributed in representative regions of China with four ancient genetic components corresponding to four different ESUs [42].

In fact, in biological systems with simple evolutionary processes, traditional genetic techniques can directly divide protection units, but it is difficult to determine clear protective units for complex evolutionary systems, such as populations with hybridization and introgression in evolutionary history [43]. Hybridization refers to interspecific hybridization between individuals from different species or populations, such as hybridization between two closely related species with bidirectional gene exchange. Introgression refers to the transfer of alleles from one species to another, and gene exchange is unidirectional. Hybridization and introgression make it more difficult to partition protective units, as analyzing different parts of the genome is likely to yield different results. Due to the impact of human activities, the displacement of organisms and habitat transfer have significantly increased the hybridization and introgression rates among various species worldwide, further increasing the threat to existing species. Genomics technology can not only effectively distinguish between natural hybridization and artificial hybridization, but also predict the impact of hybridization on species fitness (heterosis or outbreeding decline) [44].

Overall, conservation genomics serve as an efficient tool for resolution of phylogenetic relationships and elucidation of population genetic structure, population evolution history. It can help correctly define the taxonomic status of an endangered species, precisely evaluate the hybridization risk and genetic diversity, and provide valuable management information for species conservation.

3.2 Analysis of genetic diversity

Genetic diversity is closely related to the adaptive evolution and evolutionary potential of species. Traditional genetic diversity assessment is based on mitochondrial genes, microsatellites and other molecular genetic marker to calculate the genetic diversity of different populations. However, these calculations are only based on the assessment of allele frequency of certain loci represented by genetic marker, which cannot comprehensively reflect the level and panorama of genetic diversity of gene sequences in key coding regions of species. With the rapid development of genomics, it has become possible to assess genetic diversity at the whole genome level. Genomic diversity refers to the overall genetic diversity of a species or population based on variation loci at the whole genome level. In recent years, with the continuous decoding of genome sequences of endangered species and the accumulation of population genome data based on re-sequencing, more and more endangered species' genomic diversity has been evaluated.

Using genome re-sequencing method, the genetic diversity of some plant species listed in IUCN as critically endangered, endangered, or vulnerable have been studied. Generally, threatened species display lower genetic diversity than threatened and endangered species. The genetic diversity of critically endangered species endemic to China ranged from 0.0016 to 0.0030, while that of endangered species ranged from 0.0031 to 0.0038 [33].

Huang et al. (2012) sampled 446 diverse individuals of the critically endangered wild rice *Oryza rufipogon* across Asia and Oceania and sequenced them with twofold genome coverage. The sequence diversity was estimated at 0.003 which agrees with previous suggestions that as an immediate progenitor of cultivated rice, part of the genetic diversity of *O. rufipogon* was lost during domestication [45]. Ma et al. reported the first chromosome-level genome of a critically endangered species *Rhododendron griersonianum* which contributed to about 10% of horticultural rhododendron varieties. Lower genetic diversity of *R. griersonianum* (0.0019) compared to other relative species within the same genus and most other woody plants suggested that *R. griersonianum* could face large challenges to its future survival. Therefore, ex situ conservation and artificial supplementary pollination should be conducted as a priority [36].

It was also found that some living fossil plants, such as *Ginkgo biloba*, have high genetic diversity despite relatively few phenotypic variations, supporting the hypothesis of evolutionary capacitance [41].

It is worth noting that some research populations have limited sampling and insufficient representativeness, which may also lead to differences in the levels of genetic diversity reported in different endangered plants. The conservation genomics research of endangered plants in the future needs to ensure sufficient sample size in order to comprehensively evaluate the genomic diversity at the species or population level.

3.3 Inference of population dynamics history

The history of population dynamics is an important research topic in conservation genomics and population genetics [46]. As important external forces in the process of population evolution, climate change and geological events can affect the population dynamic changes. In addition, gene mutation, genetic drift, natural selection and other forces act as internal driving forces, causing the population evolution to gradually deviate from the "ideal population" and form a new pattern of genetic diversity. Therefore, studying the history of population dynamics can help us understand the shaping of population structure by climate change, geological events, and human activities, and develop reasonable management plans to cope with environmental changes [47–49]. The assessment of population dynamics in traditional genetics is mainly based on the analysis of fossil records and the history of climate or geological changes. With the development of molecular genetics, the widespread use of genetic marker provides us more comprehensive and accurate understanding about species evolution history. Molecular genetics studies based on mitochondria, chloroplasts or microsatellite genetic markers can be used to analyze the difference of polymorphism in different populations, or to infer population history dynamics by comparing the theoretical values under normal evolutionary state of populations. However, these methods require analysis of a large number of samples at the population level. On the other hand, they can only be used to trace the most recent population dynamics event instead of reflecting the overall species historical dynamics.

In the speculation of population dynamic history, the effective population size (*Ne*) is a key parameter reflecting the historical status of a population. It has important guiding significance for species conservation [50]. The effective population size (*Ne*) represents the population size under random mating conditions [51, 52]. Generally speaking, it is smaller than the actual population size. We can directly estimate the genetic drift rate of a population through *Ne*, and infer the genetic diversity and inbreeding level of the population [53]. Keeping the population size large enough to minimize the impact of genetic drift and inbreeding has always been an important goal of endangered species protection [54]. The "50/500" theoretical rule in conservation biology points out that populations with *Ne* less than 50 are vulnerable to inbreeding depression, so when isolated populations are protected alone, the population *Ne* should be kept above 50. Although occasional decrease in population size to this order of magnitude will not have an immediate adverse impact on the population, it is necessary to ensure that the population's *Ne* is above 500 in order to maintain genetic variation in the long term [55, 56].

Thus it can be seen, the *Ne* value has important guiding significance for conservation biology. The development of advanced genome technologies not only allows for more accurate estimation of population *Ne* and reconstruction of the process of population number changes of a species, but also serves as an important supplement to the field of conservation genomics.

Davidia involucrata is a rare plant endemic to China, known as the "panda among the trees". A high-quality genome of D. involucrate was constructed using PacBio Hi-C technology. Re-sequencing analysis showed that the effective population size of *D. involucrata* has been decreasing since the Quaternary Ice Age, indicating that climate change may be one of the main reasons why *D. involucrata* has become endangered. Therefore, the sensitivity of *D. involucrata* to climate change should be fully considered regarding genetic conservation of the species [57].

The effect of ice age on population size decrease can also be observed in another threatened species, *Cercidiphyllum japonicum*, which is one of the most widely distributed forest species in the Tertiary relict leaf forest of East Asia. It is listed as class II protection species according to the List of Wild Plants Under State Protection in China. Comparative genomics and population genomics research based on high-throughput sequencing technology has successfully reconstructed the evolutionary history of the Tertiary relict plant genus *Cercidiphyllum* in East Asia at the whole genome level. It found that the dry and cold environment in the middle Miocene (c. 10 Ma) led to the species differentiation of *C. japonicum* and *C. magnificum*. During the alternation of Pliocene and Pleistocene and the last glacial maximum, the population of *C. japonicum* and *C. grandiflorum* contracted sharply. However, different from the case of *D. involucrate*, selection clearance and balanced selection related to local adaptation jointly maintained genetic variations of genes involved in key physiological processes, thereby improving the ability of *C. japonicum* to adapt to different environments [58].

On the other hand, recovery of temperature during the interglacial period may be beneficial for population expansion. Genomic-based study on the population dynamics history of *Liriodendron chinense* and *Liriodendron tulipifera* showed that throughout the Quaternary ice age, the population of *L. tulipifera* continued to decrease, while the population of *L. chinense* recovered and reached its peak at approximately 0.4 mya, which is the interglacial period between the Guxiang Ice Age (0.3–0.13 million years ago) and the Nie Nie Xiongla Ice Age (0.72–0.5 million years ago). This may explain the higher genetic diversity of *L. chinense* than *L. tulipifera* [59].

Other than climate factors, geological historical events and human interference are also important reasons for population size fluctuations. Genomic conservation study of *Rhododendron griersonianum* suggested that habitat loss caused by human activities, the extremely low genetic diversity of Rhododendron rubra, and the genetic bottlenecks, inbreeding, and harmful mutations related to heat adaptation caused by geological historical events are the main reasons for the formation and maintenance of the 'extremely small population' of *R. griersonianum* [36].

In summary, whole genome sequencing can help infer fluctuations in effective populations and track historical events of population dynamics, as well as infer the impact of past geological and climatic events on the number and genetic composition of contemporary effective populations. As shown in the above research, extreme geological and climatic changes and human activities are important reasons for the rapid decline in effective population size and genetic diversity of endangered species. Linking population size changes with historical environmental changes can also help predict the impact of future environmental changes on population distribution and genetic diversity.

3.4 Inbreeding and asexual reproduction

According to Hardy–Weinberg principle, in an ideal state, individuals in the population mate randomly, and the allele frequency is stable in inheritance. However, there are generally some non-random mating patterns in natural populations, the most common of which are inbreeding and asexual reproduction, which will affect the homozygosity of the genome. The genetic differences in individual genomes within a population are the key to population adaptation and evolution. Genetic recombination will generate new gene combinations, so outbreeding between unrelated individuals can effectively increase the genetic diversity of a population. On the other hand, selfing or inbreeding between close relatives can significantly reduce the level of genetic variation within the population. Compared with the outbred individuals, the population fitness of the inbred progeny will be significantly reduced due to the accumulation of harmful mutations. This is called inbreeding depression [60]. The degree of inbreeding depression depends on the genetic load of the population. In nature, inbreeding is the result of limited population size, so some rare species are often more vulnerable to threats due to their small populations and isolation. Most individuals in their populations exhibit consistent genotypes with their ancestors [61]. The reproductive method of clone reproduction lacks genome recombination, resulting in extremely low population genetic diversity [62]. In addition, the accumulation of harmful mutations from generation to generation further reduces the adaptive potential of the species [62], leading to a "dead end" in species evolution [63].

Whole genome sequencing can provide us genome-level genetic markers to accurately estimate the inbreeding level of the population. Due to the wide coverage and low computational cost of SNPs on the genome, they are currently commonly used for phylogenetic analysis and Run of Homozygosity (ROH) analysis to evaluate inbreeding levels. Phylogenetic relationship analysis refers to the calculation of relative values of genetic similarity between individuals, which can be inferred through genotype markers. Common evaluation parameters include coefficient of kinship, co-ancestry, and identity by descent (IBD) [64]. Continuous homozygous regions on the individual genome are the result of inbreeding. When inbreeding occurs, parents will pass the same haplotype segments to their offspring to generate continuous homozygous segments. Therefore, the degree of inbreeding can be reflected by the proportion of ROH in the genome (FROH) [65, 66]. After a period of time, these fragments will be disrupted by recombination, so the length of ROH can also be used to estimate the time of inbreeding events [66].

At present, various software were developed to use genomic data to detect harmful mutations in populations. For example, SIFT prediction software based on sequence homology can help analyze whether newly emerged non synonymous mutations are harmful mutations [67]. PolyPhen-2 prediction software based on sequence homology and protein structure can help predict harmful mistranslation mutations in populations [68]. These software have been applied to the detection of inbreeding depression and harmful mutation of several endangered plant populations. For example, genomic study on two closely-related species of *Ostrya* found that the effective population sizes of the critically endangered species *Ostrya rehderiana* and the widespread species *Ostrya polynervis* both decline in the Quaternary Glacial Age. The effective population size of *Ostrya polynervis* rose rapidly after the end of Glacial Age, while the effective population size of *Ostrya rehderiana* continued to decline since the Holocene after the end of the Glacial Age, along with accumulation of harmful mutations in the genome and occasional infertility of natural fruiting. Surprisingly, the extremely harmful mutations in *Ostrya rehderiana* were significantly reduced compared to the widely distributed species *Ostrya polynervis*. Therefore, the prolonged decline in effective population size leads to weakened inbreeding inhibition, coupled with a reduction in extremely harmful mutations, allowing the endangered *Ostrya rehderiana* to remain robust and survive indefinitely [7]. Instead of simply increasing the total number of surviving individuals by collecting inbred seeds or cloning cuttings in endangered trees, artificial hybridization strategies should be designed in the future to reduce the risk of inbreeding and inheritance, and loss of diversity caused by drift transmission [7].

Based on this idea, Ma et al. analyzed the inbreeding and harmful mutation patterns of different populations of the endangered species *Acer yangbiense*, and developed personalized genetic rescue models. For example, genetic rescue is carried out for populations with high FROH and high homozygous harmful mutations. Pollination of female flowers in populations with the highest number of harmful mutations using pollens from the lowest number of homozygous harmful mutations can not only avoid introducing more harmful mutations, but also hybridize homozygous harmful mutations [69].

The above studies provide important reference value for future related research. More genomic data on endangered plants and continuously improved algorithms can help researchers and managers better identify endangered populations and understand the impact of different management methods on species fitness and genome.

3.5 Adaptive evolution

Adaptive evolution refers to the continuous accumulation of genetic variation conducive to survival and reproduction under the pressure of natural selection to better adapt to the environment and increase fitness. Traditional adaptive evolution research is mainly conducted through candidate gene methods. However, the candidate gene method relies on screening, targeted amplification, and comparison of known candidate genes. The analysis efficiency is low. The emergence of genomic data has made it possible to detect signals of adaptive evolution at the whole genome level, and has been widely applied in the study of adaptive evolution and population localization adaptation of endangered species.

Compared with only considering the genetic diversity of the whole genome, the adaptability of key regions or loci on the genome has a more significant impact on the adaptability of species. Under the neutral background of the whole genome, this part of the region will be an obvious outlier. Therefore, the possible adaptive association regions can be determined by detecting highly differentiated regions or selected regions on the population [70, 71].

In the genomic conservation study of *G. biloba*, 25 genes involved in insect and fungal defenses and responses to abiotic stress such as dehydration, low temperature and high salt, suggesting that ginkgo possessed unusually high resistance or tolerance to both abiotic and biotic stress, particularly herbivores and pathogens [41]. In *C. japonicum*, 823 genes that may be related to local adaptation were detected through F_{ST} and genotypic environmental association analysis. These genes are mainly enriched in cell development and proliferation, auxin metabolic pathway and stress response [58]. *Nostoc flagelliforme* is a first-class protected wild plant in China, mainly distributed in arid and desert areas in the north and northwest. Genome sequencing and comparative genome analysis of *Nostoc flagelliforme* showed that the expression of genes related to protection of photosynthetic apparatus, synthesis of monounsaturated fatty acids, ultraviolet radiation response promoted adaptation of *Nostoc flagelliforme* to high intensity ultraviolet radiation and extreme drought conditions. The study provides new insights into the research on the adaptability of blue-green algae to adversity [72].

In some other endangered plant species, the genomic loci under natural selection were also identified. Wang et al. used two genotypic-environmental association methods to conduct genome screen on 185 wild soybean germplasm individuals distributed in the three major agricultural ecological regions of China. Multiple genes involved in local adaptation, such as flowering time and temperature related genes were identified. A positive selection site was found on chromosome 19, which contains two adjacent MADS box transcription factors, possibly related to the ability of wild soybeans to adapt to high latitude environments [73].

In addition, through genome analysis of some endangered wild relatives of plants with economic value, we can find the impact of artificial domestication on adaptive evolution. In a genomic study of 81 cultivated and wild tea (*Camellia sinensis*) individuals, the chromosomal selection-sweeping regions were identified and enriched in the endemic and ancient species. It was found that artificial domestication not only improved the flavor of locally cultivated tea trees, but also improved their resistance to non-biotic stress [74]. Niu et al. re-sequenced 38 individual samples of *Dendrobium officinale* and its five related species from 13 regions. Combined genome-wide association studies (GWAS) identified 13 GWAS loci in total. The related genes at these loci are mostly associated with morphological traits such as plant height, leaf length, and stem length, which may be affected by artificial domestication [75].

In conclusion, genome wide sequencing is a powerful tool for detecting natural selection signals, revealing the genetic basis of phenotypic traits, and identifying local adaptation. It can promote our understanding of the inherent mechanisms of genetic variation and adaptive characteristics and help us take targeted protective measures to promote the adaptation of endangered plants to rapidly changing natural environments.

4. Future prospects

The current global rate of species extinction is accelerating, and the loss of biodiversity and ecosystem degradation has posed significant risks to human survival and development. Solving the problems of long-term survival for threatened species and constructing long-term protection mechanisms are urgently needed tasks.

Genomics study has innovated species conservation methods from multiple directions, such as identification of lineages and inbreeding events, identification of adaptive loci and outbreeding decline loci based on a large number of genetic markers. Conservation genomics can not only solve scientific problems such as genetic diversity, population genetic structure, and population dynamics that traditional conservation genetics focuses on, but also further trace the evolutionary history of species from ancient times to the present, and analyze the molecular mechanisms of population local adaptation and species adaptive evolution. In addition, conservation genomics also provides the possibility to study the genetic basis of interspecific interactions, promoting population level management rather than individual-level protection.

The classic approach in genomics study on endangered plants mainly include the following pipelines: constructing high-quality genomes, conducting comparative genome analysis to identify genes that expand and contract, elucidating the evolutionary process of genomes through WGD analysis or repeat sequence analysis. At the same time, it can further analyze and verify the gene family related to special phenotype and special mechanism by combining with transcriptome analysis. Together with genome re-sequencing, the population historical dynamics can be reveled and valuable genes can be found for genetic breeding.

At the same time, it should also be noticed that although genomics study can provide preliminary solutions for biodiversity conservation, there are still some limitations. Firstly, in order to truly achieve endangered species protection and resource value development, more genomic information of endangered species should be obtained. The techniques of artificial pollination, hybridization, and expansion of endangered plants should also be tackled. Secondly, the vast majority of comparative genomics and population genomics research is currently based on the analysis of single nucleotide polymorphism (SNP) markers, while little is known about the role of structural variations (SV) in the adaptive evolution and local adaptation of endangered species. With the widespread application of third-generation sequencing technology, the ability to accurately analyze SV has greatly improved, which will undoubtedly promote understanding of the role of SV in the adaptive evolution of endangered species and local population adaptation. Finally, although genomics technology has updated some of our scientific knowledge in the field of conservation biology, it may not necessarily be useful for species conservation. For example, based on genomics analysis, we can interpret the function of genetic variations at the genome level and their impact on individual fitness. However, it is still not enough for us to evaluate the survival ability of the population, unless we can link individual fitness with population growth rate for discussion [76]. And this requires long-term research on individual fitness and its impact on population growth rate. This may be the biggest challenge currently facing conservation genomics.

Acknowledgements

We would like to thank the Zhejiang Provincial Welfare Technology Applied Research Project [LGN21C020007], the National Natural Science Foundation of China [No. 31800187], the 2021 and 2022 Foreign Specialized Projects of the Ministry of Science and Technology [QN2021016002L, G2022016022L].

Conflict of interest

The authors declare no conflict of interest.

Author details

Qing Ma[1], Gang Wu[2], Wenjie Li[1], Seyit Yuzuak[3], Fachun Guan[4] and Yin Lu[1*]

1 College of Biology and Environmental Engineering, Zhejiang Shuren University, Hangzhou, Zhejiang, China

2 Institute of Science and Technology and Culture, Zhejiang University of Science and Technology, Hangzhou, Zhejiang, China

3 Department of Molecular Biology and Genetic, Burdur Mehmet Akif Ersoy University, Burdur, Turkey

4 Institute of Rural Energy and Ecology, Jilin Academy of Agricultural Sciences, Changchun, China

*Address all correspondence to: maqing@zjsru.edu.cn; luyin@zjsru.edu.cn

References

[1] Haddad NM, Brudvig LA, Clobert J, et al. Habitat fragmentation and its lasting impact on Earth's ecosystems. Science Advances. 2015;**1**:e500052

[2] Newbold T, Hudson LN, Hill SL, et al. Global effects of land use on local terrestrial biodiversity. Nature. 2015;**520**:45-50

[3] Cronk Q. Plant extinctions take time. Science. 2016;**353**:446-447

[4] Miraldo A, Li S, Borregaard MK, et al. An Anthropocene map of genetic diversity. Science. 2016;**353**:1532-1535

[5] UK, R.B.G. State of the world's plants report-2016. 2016

[6] Xu S, He Z, Zhang Z, et al. The origin, diversification and adaptation of a major mangrove clade (Rhizophoreae). National Science Review. 2017;**4**:721-734

[7] Yang Y, Ma T, Wang Z, et al. Genomic effects of population collapse in a critically endangered ironwood tree Ostryarehderiana. Nature Communications. 2018;**9**:1-9

[8] DeSalle R, Amato G. Conservation Genetics in the Age of Genomics. New York, USA: Columbia University Press; 2009. pp. 1-24

[9] Avise JC. Perspective: Conservation genetics enters the genomics era. Conservation Genetics. 2010;**11**: 665-669

[10] Allendorf FW. Genetics and the conservation of natural populations: Allozymes to genomes. Molecular Ecology. 2017;**26**:420-430

[11] Allendorf FW, Hohenlohe PA, Luikart G. Genomics and the future of conservation genetics. Nature Reviews Genetics. 2010;**11**:697-709

[12] Myers N. Threatened biotas: "Hot spots" in tropical forests. Environmentalist. 1988;**8**:187-208. DOI: 10.1007/BF02240252

[13] Barnosky AD, Matzke N, Tomiya S, et al. Has the Earth's sixth mass extinction already arrived? Nature. 2011;**471**:51-57

[14] Ceballos G, Ehrlich PR, Barnosky AD, et al. Accelerated modern human–induced species losses: Entering the sixth mass extinction. Science Advances. 2015;**1**:e400253

[15] IUCN. The IUCN Red List of Threatened Species. Version 2022-2. 2023. Available from: https://www.iucnredlist.org

[16] Fuentes-Pardo AP, Ruzzante DE. Whole-genome sequencing approaches for conservation biology: Advantages, limitations and practical recommendations. Molecular Ecology. 2017;**26**:5369-5406

[17] Hohenlohe PA, Funk WC, Rajora OP. Populationgenomics for wildlife conservation and management. Molecular Ecology. 2021;**30**:62-82

[18] Vasemägi A, Primmer CR. Challenges for identifying functionally important genetic variation: The promise of combining complementary research strategies. Molecular Ecology. 2005;**14**:3623-3642

[19] Sanger F, Nicklen S, Coulson AR. DNA sequencing with chain-terminating

inhibitors. Proceedings of the National Academy of Sciences. 1977;**74**:5463-5467

[20] Mardis ER. Next-generation DNA sequencing methods. Annual Review of Genomics and Human Genetics. 2008;**9**:387-402

[21] Roberts RJ, Carneiro MO, Schatz MC. The advantages of SMRT sequencing. Genome Biology. 2013;**14**:405

[22] Loman NJ, Quick J, Simpson JT. A complete bacterial genome assembled de novo using only nanopore sequencing data. Nature Methods. 2015;**12**:733-735

[23] Lee H, Gurtowski J, Yoo S, et al. Third-generation sequencing and the future of genomics. BioRxiv. 2016;**2016**:048603. DOI: 10.1101/048603

[24] Wang GT, Wang ZF, Wang RJ, et al. Development of microsatellite markers for a monotypic and globally endangered species, *Glyptostrobus pensilis* (Cupressaceae). Applications in Plant Sciences. 2019;**7**:e01217

[25] Bao WQ, Ao D, Wuyun T, et al. Development of 85 SNP markers for the endangered plant species *Prunus mira* (Rosaceae) based on restriction site-associated DNA sequencing (RAD-seq). Conservation Genetics Resources. 2020;**12**:525-527

[26] Cai CN, Xiao JH, Ci XQ, et al. Genetic diversity of *Horsfieldia tetratepala* (Myristicaceae), an endangered plant species with extremely small populations to China: Implications for its conservation. Plant Systematics and Evolution. 2021;**307**:50

[27] Andrews KR, Good JM, Miller MR, et al. Harnessing the power of RADseq for ecological and evolutionary genomics. Nature Reviews Genetics. 2016;**17**:81-92

[28] Qi YX, Liu YB, Rong WH. RNA-Seq and its applications: A new technology for transcriptomics. Hereditas (Beijing). 2011;**33**:1191-1202 (in Chinese with English abstract)

[29] Warr A, Robert C, Hume D, et al. Exome sequencing: Current and future perspectives. 3G (Genes, Genomes, Genetics). 2015;**5**:1543-1550

[30] Chong ZC, Ruan J, Wu CI. Rainbow: An integrated tool for efficient clustering and assembling RAD-seq reads. Bioinformatics. 2012;**28**:2732-2737

[31] Eaton DAR. PyRAD: Assembly of de novo RADseq loci for phylogenetic analyses. Bioinformatics. 2014;**30**:1844-1849

[32] Jones MR, Good JM. Targeted capture in evolutionary and ecological genomics. Molecular Ecology. 2016;**25**:185-202

[33] Jing ZY, Cheng KG, Shu H, et al. Whole genome resequencing approach for conservation biology of endangered plants. Biodiversity Science. 2023;**31**(5):22679

[34] Lewin HA, Robinson GE, Kress WJ, et al. Earth BioGenome Project: Sequencing life for the future of life. Proceedings of the National Academy of Sciences, USA. 2018;**115**:4325-4333

[35] Frankham R. Where are we in conservation genetics and where do we need to go? Conservation Genetics. 2010;**11**:661-663

[36] Ma H, Liu YB, Liu DT, et al. Chromosome-level genome assembly and population genetic analysis of a critically endangered Rhododendron provide insights into its conservation. The Plant Journal. 2021;**107**:1533-1545

[37] Rossetto M, Yap J, Lemmon J, et al. A conservation genomics workflow to guide practical management actions. Global Ecology and Conservation. 2021;**26**:e01492

[38] Waples RS. Definition of "species" under the Endangered Species Act: Application to Pacific salmon. US Government Technology Report. 1991;**53**:3

[39] Frunk WC, McKay JK, Hohenlohe PA, et al. Harnessing genomics for delineating conservation units. Trends in Ecology & Evolution. 2012;**27**:489-496

[40] Forester BR, Murphy M, Mellison C, et al. Genomics-informed delineation of conservation units in a desert amphibian. Molecular Ecology. 2011;**31**:5249-5269

[41] Zhao YP, Fan GY, Yin PP, et al. Resequencing 545 ginkgo genomes across the world reveals the evolutionary history of the living fossil. Nature Communications. 2019;**10**:4201

[42] Liu PL, Zhang X, Mao JF, et al. The Tetracentron genome provides insight into the early evolution of eudicots and the formation of vessel elements. Genome Biology. 2020;**21**:291

[43] Waples RS, Kays R, Fredrickson RJ, et al. Is the red wolf a listable unit under the US Endangered Species Act? Journal of Heredity. 2018;**109**:585-597

[44] Allendorf FW, Leary RF, Spruell P, et al. The problems with hybrids: Setting conservation guidelines. Trends in Ecology & Evolution. 2001;**16**:613-622

[45] Huang XH, Kurata N, Wei XH, et al. A map of rice genome variation reveals the origin of cultivated rice. Nature. 2012;**490**:497-501

[46] Girod C, Vitalis R, Leblois R, et al. Inferring population decline and expansion from microsatellite data: A simulation-based evaluation of the Msvar method. Genetics. 2011;**188**:165-179

[47] Elmer KR, Reggio C, Wirth T, et al. Pleistocene desiccation in East Africa bottlenecked but did not extirpate the adaptive radiation of Lake Victoria haplochromine cichlid fishes. Proceedings of the National Academy of Sciences. 2009;**106**:13404-13409

[48] Franks SJ, Kane NC, O'Hara NB, et al. Rapid genome-wide evolution in Brassica rapa populations following drought revealed by sequencing of ancestral and descendant gene pools. Molecular Ecology. 2016;**25**:3622-3631

[49] Fahrig L. Ecological responses to habitat fragmentation per se. Annual Review of Ecology, Evolution, and Systematics. 2017;**48**:1-23

[50] Wang J, Santiago E, Caballero A. Prediction and estimation of effective population size. Heredity. 2016;**117**:193-206

[51] Wright S. Evolution in Mendelian populations. Genetics. 1931;**16**:97

[52] Wright S. Inbreeding and homozygosis. Proceedings of the National Academy of Sciences of the United States of America. 1933;**19**:411

[53] Kempthorne O. Evolution and the genetics of populations. Volume 2: The Theory of Gene Frequencies. JSTOR. 1971;**1971**:120-121

[54] McElhany P, Rucklelshaus MH, Ford MJ, et al. Viable Salmonid Populations and The Recovery of Evolutionarily Significant Units. U.S. Dept. Commer. NOAA Tech. Memo. NMFS-NWFSC.; 2000. pp. 42-156

[55] Franklin IR. Evolutionary changes in small populations. Conservation Biology. 1980;**1980**

[56] Jamieson IG, Allendorf FW. How does the 50/500 rule apply to MVPs? Trends in Ecology & Evolution. 2012;**27**:578-584

[57] Chen Z, Ai F, Zhang J, et al. Survival in the tropics despite isolation, inbreeding and asexual reproduction: Insights from the genome of the world's southernmost poplar (*Populus ilicifolia*). The Plant Journal. 2020;**103**:430-442

[58] Zhu SS, Chen J, Zhao J, et al. Genomic insights on the contribution of balancing selection and local adaptation to the long-term survival of a widespread living fossil tree, *Cercidiphyllum japonicum*. New Phytologist. 2020;**228**:1674-1689

[59] Chen J, Hao Z, Guang X, et al. Liriodendron genome sheds light on angiosperm phylogeny and species-pair differentiation. Nature Plants. 2019;**5**:18

[60] Glémin S, Bataillon T, Ronfort J, et al. Inbreeding depression in small populations of self-incompatible plants. Genetics. 2001;**159**:1217-1229

[61] Frankham R. Inbreeding and extinction: A threshold effect. Conservation Biology. 1995;**9**:792-799

[62] McKey D, Elias M, Pujol B, et al. The evolutionary ecology of clonally propagated domesticated plants. New Phytologist. 2010;**186**:318-332

[63] Mattila TM, Laenen B, Slotte T. Population genomics of transitions to selfing in Brassicaceae model systems. In: Statistical Population Genomics. New York, NY: Humana; 2020. pp. 269-287

[64] Visscher PM, Medland SE, Ferreira M, et al. Assumption-free estimation of heritability from genome-wide identity-by-descent sharing between full siblings. PLoS Genetics. 2006;**2**:e41

[65] Lencz T, Lambert C, DeRosse P, et al. Runs of homozygosity reveal highly penetrant recessive loci in schizophrenia. Proceedings of the National Academy of Sciences. 2007;**2007**(104):19942-19947

[66] McQuillan R, Leutenegger AL, AbdelRahman R, et al. Runs of homozygosity in European populations. The American Journal of Human Genetics. 2008;**83**:359-372

[67] Ng PC, Henikoff S. SIFT: Predicting amino acid changes that affect protein function. Nucleic Acids Research. 2003;**31**:3812-3814

[68] Adzhubei IA, Schmidt S, Peshkin L, et al. A method and server for predicting damaging misense mutations. Nature Methods. 2010, 2010;**7**:248-249

[69] Ma YP, Liu DT, Wariss HM, et al. Demographic history and identification of threats revealed by population genomic analysis provide insights into conservation for an endangered maple. Molecular Ecology. 2022;**31**:767-779

[70] Nielsen R. Molecular signatures of natural selection. Annual Review of Genetics. 2005;**39**:197-218

[71] Oleksyk TK, Smith MW, O'Brien SJ. Genome-wide scans for footprints of natural selection. Philosophical Transactions of the Royal Society B Biological Sciences. 2010;**365**:185-205

[72] Shang JL, Chen M, Hou S, et al. Genomic and transcriptomic insights into the survival of the subaerial cyanobacterium Nostoc flagelliforme in arid and exposed habitats.

Environmental Microbiology. 2019;**21**(2):845-863

[73] Wang J, Hu ZB, Liao XL, et al. Whole-genome resequencing reveals signature of local adaptation and divergence in wild soybean. Evolutionary Applications. 2022;**15**:1820-1833

[74] Xia E, Tong W, Hou Y, An Y, et al. The reference genome of tea plant and resequencing of 81 diverse accessions provide insights into genome evolution and adaptation of tea plants. Molecular Plant. 2020;**13**:14

[75] Niu ZT, Zhu F, Fan YJ, et al. The chromosome-level reference genome assembly for *Dendrobium officinale* and its utility of functional genomics research and molecular breeding study. Acta Pharmaceutica Sinica B. 2021;**11**:2080-2092

[76] Coulson T, Benton TG, Lundberg P, et al. Estimating individual contributions to population growth: Evolutionary fitness in ecological time. Proceedings Biological Sciences. 2006;**273**:547-555

Chapter 5

Dragon's Blood Tree (*Dracaena cinnabari*): A Cenozoic Relict

Sanjay Saraf

Abstract

The Dragon's blood tree (*Dracaena cinnabari*) woodland is one of the oldest surviving endangered forest communities on Earth. This unique endemic species of Dragon's blood tree is famous since antiquity for its bright red resin "Dragon's blood" and umbrella-shaped canopy. They are almost extinct everywhere except present as small habitats in Socotra Archipelago (Yemen), a UNESCO World Heritage Site. In the last two decades, there has been a significant decline in Dragon's blood tree population in the archipelago, posing a threat to its existence. We attempt to review the status of Dragon's blood tree population in Socotra Archipelago, factors affecting its survival, and the status of conservation efforts propose recommendations to preserve this flagship species.

Keywords: *Dracaena cinnabari*, Dragon's blood tree, Socotra Island, biodiversity, vulnerable, endemism

1. Introduction

The Socotra Archipelago is one of the most significant and well-preserved island ecosystems in the world and holds global significance for the exceptional level of biodiversity and endemism in its ecosystem. Socotra Island is a masterpiece of evolution, containing a unique assemblage of species and habitats, and is ranked as the world's tenth richest island for endemic plant species per square kilometer, with 37% endemic species, which makes the ratio comparable with that of the Galapagos and higher than those found in Mauritius, Rodrigues, or the Canary Islands [1]. The high level of endemism seen in the archipelago is also in accordance with its estimated geological age and ecosystem. Socotra island is home to over 350 species of endemic plant species and is considered as a gem of biodiversity in the Arabian Sea [2–4]. Dragon's blood tree significantly contributes to this endemic biodiversity and is the dominant iconic species of the island (**Figure 1**), (Video available at https://bit.ly/3Dx92h6), [5].

1.1 Dragon's blood tree geographical location

The Socotra Archipelago is part of the Republic of Yemen and is geographically located in the north-western corner of the western Indian Ocean, at the junction

Figure 1.
Dragon's blood tree (D. cinnabari), "flagship species of Socotra" (photograph by the author, Socotra 2018).

Figure 2.
Overview map of the Socotra archipelago, a remarkable biodiversity spot, showing its isolated geographic location in the Indian Ocean.

between the Gulf of Aden and the Arabian Sea, about 380 km off the coastline of Yemen and 250 km east of Cape Guardafui (Somalia), the tip of the Horn of Africa (**Figure 2**).

The Socotra Archipelago is 250 km long and comprises a group of four islands and other small islets. Socotra forms the largest island of the eponymous archipelago of four islands (about 95% of its landmass), followed by the three satellite islands,

Figure 3.
Socotra archipelago "Galapagos of the Indian Ocean" long geologic isolation is the key to preservation of Socotra's endemic ecosystem.

which are collectively known as "the brothers," Samha, Darsa, and Abd Al Kuri, and other small rocky islets Jazirat, Sabuniya, and Ka'l Fir'awn, which are uninhabitable (**Figure 3**) but exhibits high level of endemism, rendering the archipelago as a whole even more significant.

The official name of the island is Socotra (often spelled as sokotra, Soqotra or Suqotra), which emanated from the Arabic word *Souq-Al-Qatr,* 'Souq' meaning "the market" and 'Cotra' referring to "dragon's blood"—a reference to the bright red resin produced by Dragon's blood tree species. The main language is Socotri, which is one of the Semitic languages, and Arabic is the official and commonly spoken language.

1.2 Dragon's blood tree taxonomic hierarchy

Dracaena (Greek: δράκαινα, drakaina "female dragon") genus comprises between 60 and 100 Dracaena species, *D. cinnabari* being one of the only six species that grow as a tree and is considered a terrestrial monocotyledon representative of the Tertiary flora (**Table 1**, **Figure 4**).

1.3 Dragon's blood tree ecosystem evolution and morphological characteristics

Dragon's blood trees species belong to one of the oldest ecosystems in the world. Dragon's blood tree species are spectacular relicts of the Mio-Pliocene Laurasian subtropical forest in Socotra (Yemen) [7]. In this epoch, Dragon's tree vegetation extended in a continuous vegetation belt between Northern Africa and Southern Europe, but afterward, it was disrupted due to climatic changes, causing the desertification of North Africa [8]. Today, *D. cinnabari* is vulnerable and almost extinct everywhere due to the Pliocene climate changes and extensive desertification of North Africa and Southern Europe [4, 7, 9].

D. cinnabari is a monocotyledonous tree with a distinctive umbrella-shaped canopy due to a "dracoid" ramification of branches (**Figure 5**), [10]. Dracaena species are exceptional among monocotyledonous plants because of their capacity for

Figure 4.
Dragon's blood tree (D. cinnabari Baif.f): 'Singular dominant endemic community of Socotra.' This unique tree symbolizes a close bond between nature and the indigenous inhabitants of the island. (photograph by the author, Socotra 2018).

Kingdom	Plantae
Subkingdom	Viridiplantae
Infrakingdom	Streptophyta
Superdivision	Embryophyta
Division	Tracheophyta
Subdivision	Spermatophytina
Class	Magnoliopsida
Superorder	Lilianae
Order	Asparagales
Family	Asparagaceae
Subfamily	Nolinoideae
Genus	Dracaena L.
Species	*D. cinnabari* Baif.f Dragon's Blood Tree

Table 1.
Taxonomic serial No.: 505865 [6].

secondary thickening of stems and roots [11]. The *D. cinnabari* is a dominant endemic evergreen tree and can live for more than 500 years, often reaching a height of 35 to 39 feet and is vulnerable to extinction.

1.4 Dragon's blood tree present distribution status

The Socotra Island is approximately 110 km long and about 40 km wide, with a total surface area of 3625 km^2. The island can be geographically divided into three main zones, namely Haggier (Hagghier, Hajhir) mountains, limestone plateau with

Figure 5.
Dragon's blood trees (D. cinnabari): The "dracoid" ramification of branches is an adaptation to its harsh environment, which helps in capturing the moisture in the arid environment. ("physiologic plasticity") (photograph by the author, Socotra 2018).

many cave systems, and the alluvial coastal plains. The Haggier mountains form the backbone of the island and are situated in the northwest part of the island. Its highest elevation is at Jabal Dryet (1526 m) in the central Haggier Massif. In general, the summit of the limestone plateau is covered with sparse shrubland or low woody-herb communities. The coastal plains are sub desertic with open shrubland or in some areas devoid of vegetation.

In general, Dragon's blood trees occur typically in small dense groups and are found on higher slopes of the limestone mountains, particularly in the central and eastern parts of the island (**Figures 6** and 7).

Dragon's blood tree woodlands are preponderant in the large central plateau of Diksam (Dixam), the central granite massive of Haggier, and the eastern area of Hamadero, Sirahon, and Kilisan [12]. Several smaller and disrupted Dracaena populations exist on Kilim, Sirahan, Shibhon, and other less accessible localities. A dense Dragon's blood tree forest is present in the limestone plateau, known as Roqeb di Firmihin. According to a recent study [5], this small plateau occupies only 2% of the total suitable area currently occupied by the endemic Dragon's blood trees of Socotra, yet it hosts more than 40% of the living population of all *D. cinnabari* trees, making it an attractive hotspot for researchers.

Dragon's blood tree density is not homogenous and has a fragmented distribution, with predominant presence in the central and eastern parts of the island as mentioned earlier. The area of distribution ranges from an altitude of 150 to 1600 m above sea level, and it dominates above 600 meters above sea level (m.a.s.l) [5]. Dragon's blood trees are not seen in the seaside plains, lowlands (below 180 m.a.s.l), and the western part of the island [12].

In the past, Dragon's blood tree habitats were present over larger areas of the island. However, currently, habitats of Dragon's blood tree are dwindling, and several authors have described the *D. cinnabari* habitat decline on Socotra Island [1, 7, 10, 11, 13, 14].

Figure 6.
Socotra: General view of habitat. D. cinnabari Is the unique identity of Socotra Island and now occupies only 5% of its habitat. (photograph by the author, Socotra 2018).

Figure 7.
Dragon's blood tree on a higher plateau. The endangered Dragon's blood trees are strongly tied to the culture of the Socotran people. (photograph by the author, Socotra 2018).

Till the 19th century, the geology of the Socotra Archipelago received little attention, with limited mention of the Dragon's blood tree population in the Socotra archipelago. Most articles published in the late 19th and early 20th centuries were descriptions of endemic species collected by visitors and researchers to the islands. The first description of Dragon's blood tree in the literature was mentioned during the survey of Socotra led by Lieutenant J.R. Wellsted of the East India Company in 1835 [15]. He named the tree as Pterocarpus draco (Greek: πτερον, pteron, "wing" + Latin" carpus from Greek: καρπός, karpos, "fruit"; Latin: dracō from Greek: δράκων, drakōn, "dragon"). In the scientific literature, Dragon's blood tree species was first described by the Scottish botanist Sir Isaac Bayley Balfour in 1888 [16].

The best contemporary distribution of Dracaena forests and woodlands was published by Král and Pavliš in 2006 [17] using remote sensing data. They found that the habitats hosting *D. cinnabari* comprised a total of 7230 ha (hectare), including only 230 ha of Dracaena forests and 800 ha of mixed mountains forests, with the rest of the area (6200 ha) consisting of woodlands with low tree densities and overmatured populations [18, 19]. Based on statistical analyses as well as on direct field observations, Král and Pavliš also commented that the Dracaena populations on Socotra do not regenerate to a great extent, and their age structure indicates over maturity.

Attorre et al. in 2007 used a deterministic regression tree analysis (RTA) model to examine environmental variables related to the current *D. cinnabari* species distribution. They found that the current distribution and abundance of *D. cinnabari* is correlated to three factors: moisture index (i.e., the ratio between the annual precipitation and potential evapotranspiration), mean annual temperature, and slope. According to this model, *D. cinnabari* occupies only 5% of its current potential habitat, and this potential habitat is expected to be reduced by 45% by 2080 because of a predicted climate change, with increased aridity [10].

A study by Madera et al. in 2019 using remote sensing data estimated the population size of *D. cinnabari* to be 80,134 individuals, with sub-populations varying from 14 to 32,196 individuals, with an extinction time ranging from 31 to 564 years (**Figure 8**) [5].

A toponym study by Al-Okaishi (2021), carried out in four areas on Socotra Archipelago (301 toponyms), assumed that dragon's blood trees had a wider

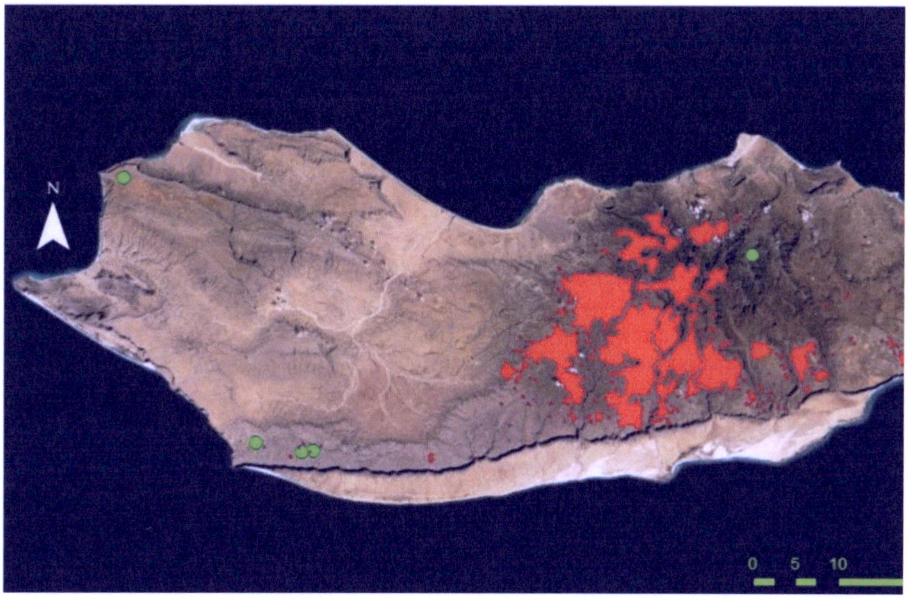

Figure 8.
The map of the distribution of the toponyms related to the D. cinnabari tree (green circles) in Socotra Island, in red, the current distribution of D. cinnabari by Maděra et al. [18].

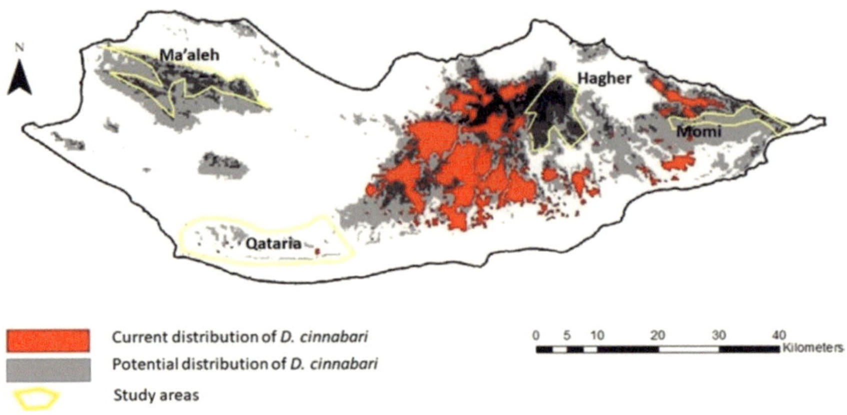

Figure 9.
Map showing the study areas (Hajher, Momi, Qatanin, Ma'aleh) in integrating two maps with the current and potential distribution of D. cinnabari according to Maděra et al. [5] and Attorre et al. [10], respectively.

distribution on Socotra Island in the past, potentially from the west in Ma'aleh to the east in Momi, before humans inhabited the island (**Figure 9**), [20].

In 2021, Vahalík et al. (published 2023) did a field survey of Socotra using a pair of UAVs (using the DJI Mavic Mini drones) to spatially describe individual tree positions, tree density, mortality, and the forest age structure. They found that the spatial distribution of the Dracaena tree density within the entire plateau is variable. The mean age of the forest, based on crown age (derived from crown size), was estimated at an average of ca. 300 years (291.5 years), with some individuals older than 500 years [21].

1.5 Dragon's blood tree conservation status

Due to its remarkable and highly vulnerable island ecosystems containing many endemic species, Socotra Archipelago was designated as a UNESCO Man and Biosphere (MAB) Reserve in 2003, a Ramsar Site in 2007 (Detwah Lagoon), and then as a UNESCO Natural World Heritage Site in 2008.

Dragon's blood tree (*D. cinnabari*) is categorized as "vulnerable" species on the IUCN Red List of Threatened Species (**Figure 10**), [22].

1.6 Dragon's blood (resin) and its significance

Since antiquity, Socotra Island was famous for its Dragon's blood, which is a remarkable bright red colored resin produced by *D. cinnabari*.

The name Dragon comes from the unique, red-colored sap or resin/latex that the tree produces. The tree is known locally as "Ahrieb" "إعريهب" and its resin "dum al-akhawin" "دم الأخوين," while derived (mixed-cooked) products are called "eda'a" "إيدع," while regionally different names can be found (**Figure 11**), [23].

Local legends say that the Dragon's blood tree (brother's blood tree) first grew on the spot where two brothers, Darsa and Samha, fought to death. The Dragon's blood (red resin) is mentioned in early literature by an unknown author of the Periplus of the Erythrean Sea around the mid-first century AD, who called it "cinnabar," possibly because of the matching color [24].

Dioscorides in 90 AD mentioned the Dragon's blood resin in his book "On Medical Material" as Kinnabari "cinnabari," brought from Africa [25]. Names of Dragon's

Figure 10.
IUCN red list aims to convey the urgency of conservation issues to the public and policy makers, as well as help the international community to reduce species extinction.

Figure 11.
The Socotra ecosystem and D. cinnabari species are unique to the island, with high global significance. (photograph by the author, Socotra 2018).

blood tree and its resin are also mentioned in old Arabic literature by travelers and researchers who visited Socotra in earlier days [26–29].

This highly prized resin has been historically harvested by the indigenous population for local use and trade throughout medieval periods for diverse medical, artistic,

Figure 12.
The Dragon's blood is a common name of a red sap, or resin, produced by the Dragon's blood tree in response to mechanical trauma. (photograph by the author, Socotra 2018).

and magical uses. It was frequently used as a medicine for respiratory and gastro-intestinal problems in the Mediterranean basin and by early Greeks, Romans, and Arabs [30–32].

Miller and Morris mention the use of *Dracaena* resin as a coloring matter for varnishes, tinctures, toothpastes, and plaster for dying the horn to make it look like tortoiseshell [1].

Dragon's "blood" secretion (**Figure 12**) can be considered an induced natural defense mechanism following trauma by cells of the stem, and during the process of wound repair, this coats the margins of the wound providing additional protection against desiccation, but unfortunately, it also makes the species vulnerable to human exploitation.

Dragon's blood has astringent effects and is frequently used as a hemostatic and antidiarrheal medicine. Though the biological basis for its secretion and phytochemistry is still not completely known, the resin is believed by some authors to have antiviral, antibacterial, antifungal, antioxidant (flavonoids), and anti-carcinogenic properties [33–36]. Local inhabitants still use the red resin for treating diarrhea, fever, mouth ulcers; to stop bleeding; for wound healing, skin diseases, coloring material for dye, varnish, cosmetic, incense, painting, decorating earthen pots, folk music, alchemy, and performing social rituals. However, the efficacy for human use remains unsubstantiated and needs further bioassay-guided spectroscopic studies and scientific trials for establishing human safety and use.

Presently, this resin is an important product for the local communities and is the most important local nontimber forest product (NTFP). It is a source of income for the rural population in Socotra and is becoming even more important with the increasing population, unemployment, and tourism-related activities [20].

1.7 Dragon blood tree mounting challenges and extinction risk

The factors threatening the survival of Dragon's blood tree population envisage multiple reasons including overgrazing by the increasing population of livestock [10, 37]; habitat loss with insufficient regeneration of Dracaena growth; soil erosion;

Figure 13.
Dragon blood trees (D. cinnabari)—Socotra's most iconic plant. (photograph by the author, Socotra 2018).

Figure 14.
Uprooted Dragon's blood tree (photo by author, Socotra 2018).

increased aridity [10]; effect of past cyclones, namely Chapala and Megh (2015) and Mekunu (2018); climatic changes due to global warming; unsustainable human interference, which are not only rendering the Dracaena woodlands vulnerable to extinction [9] but also making the fragile biological hotspot vulnerable to desertification (**Figures 13** and **14**).

2. Discussion

The Archipelago of Socotra remained inaccessible for centuries due to its remote geographic location. This prolonged period of geological isolation of the archipelago, a complex geopolitical landscape, and variable climatic conditions with an arid ecosystem contributed to the maintenance of Socotra's distinctly rich biodiversity, with the preservation of many insular species like Dragon's blood tree for centuries.

However, the last few decades have threatened the fragile Socotra ecosystem due to multiple factors including woodland fragmentation, senescent dragon tree population, unsustainable harvesting of dragon's blood, unsustainable overgrazing, unsustainable livestock management, commercial collection of wood, introduction of invasive alien species, smuggling out of endangered Dragon's blood tree, climatic threats, uncontrolled infrastructure development including roads to the mountains, increased unsustainable tourism, lack of financial resources, non-diligent enforcement of international and national policies for bio-cultural preservation, current political instability, and post Covid-19 economic challenges.

The cattle (specially by goats) overgrazing of the vegetation including Dragon tree seeds is an important factor threatening the survival of the Dragon tree [1]. Consumption of seedlings and new sprouts, if not protected from goat and other livestock, prevents *D. cinnabari* and other species from regeneration as they have extremely low survival capacity in open habitats [10, 11]. Overgrazing also provokes soil erosion by causing loss of perennial vegetation layer and the thin organic topsoil following rains. The decline of D. cinnabari is likely to negatively affect plant diversity, reduce the abundance of rare endemic plants, and lead to homogenization of the vegetation [38]. The increasing demand of the red resin has resulted in the overexploitation of Dragon's blood tree and is one of the crucial factors adversely affecting the Dracaena population. The unsustainable traditional method of harvesting dragon's blood further compounds the deleterious effect on the Dracaena population [20]. Multiple cuts on tree to harvest resin invariably makes trees weak and vulnerable to uprooting in intense winds.

Another concerning factor is that most Dragon's blood trees on Socotra Island are senescent, with increasing mean population age and are subject to progressive Dracaena population decline due to limited natural regeneration of the species (**Figure 15**).

The loss of each Dragon's blood tree leads to a decrease in biodiversity, as Dragon's blood trees are important nurse tree [38]. Furthermore, the occurrence of a wide range of insect species depends on *D. cinnabari* [39]. Moreover, dragon's blood tree woodlands function as cloud forests, catching water from horizontal precipitation, fog, drizzle, and mist, playing a significant role in the hydrology of the island [40]. A decline in the Dragon's blood tree population density leads to land aridification, soil erosion, and desertification [41]. Given its ecological importance, *D. cinnabari* has been identified as an umbrella species of Socotra Island, with its conservation essential to preserve the island's native biota [10, 18, 39–43].

Socotra has a dry arid climate, with bimodal distribution of rainfall. The climate is dependent on the seasonal migration of the Inter tropical Convergence Zone (ITCZ) and related monsoon cycles [44]. Each year, from June until September, the seasonal Southwest monsoon blow from Africa brings hot, strong, and dry winds and occasional rainfall into Haggier mountains in Socotra. The Northeast monsoon in winter (November–January) is less pronounced and coincides with the rainy season in the north. The annual rainfall ranges between 200 mm in the coastal plains and 1000 mm

Figure 15.
Senescent Dragon's blood tree (photograph by the author, Socotra 2018).

in the high mountains [37]. The alternance of extreme desiccation and mist, brought about by the seasonal wind, had an important effect on the evolution of habitats and vegetation of the Socotra Island.

The climatic changes linked to global effects due to global warming are adding new challenges on the resilience of the vulnerable ecosystem. The resultant unreliable, irregular, and patchy monsoons with mean annual precipitation ranging from 207 to 569 mm is negatively influencing the survival of the present population of *D. cinnabari*. This is not only threatening Dragon's blood tree populations but also endangering other endemic species on Socotra Island. Global warming is perceived as a serious threat to the biodiversity of such hotspots because it is likely to exacerbate both grazing and prolonged drought periods due to unreliable monsoons making it exceedingly difficult for the recovery of vulnerable species [45, 46].

Increasing aridity due to ongoing climatic change is also negatively affecting the potential habitat of the Dragon's blood tree. The loss of Dragon's blood trees is also affecting the hydrological cycle as these plants capture horizontal precipitation [47].

There is also evidence that in the past, a traditional agricultural land use with protected wall system was prevalent on Socotra for organized farming activities for frankincense, myrrh, and dragon's blood and harvesting of aloe juice [48]. This strong traditional land-use management practices employed by the indigenous population served to protect the vegetation and biodiversity of the Socotra Archipelago, which is now lacking.

Figure 16.
Asphalt roads to mountains causing habitat fragmentation and ecosystem destruction. The future road works must minimize environmental impacts on the ecosystem. (photograph by the author, Socotra 2018).

The strict conservation of vulnerable areas included in the Socotra Archipelago Master Plan envisioned in 2002 are now not being enforced on Socotra due to administrative issues and a complicated land tenure system based on the tribal organization of society [5]. The lack of EIA enforcement capacity with deficient project planning and construction of new infrastructure including housing and several hundreds of kilometers of asphalt roads to mountains (since 2003) are also adversely affecting the fragile habitat of Socotra (**Figure 16**).

Other major threats are the increase in tourism and recreational activities, smuggling of Dragon blood trees to sell them in international markets, increasing immigration, import of goods from mainland Yemen, and pollution by deficient waste management practices around human settlements [20].

The archipelago's remarkable integrity and Outstanding Universal Value (OUV) is also significantly threatened by the unsustainable developments. There are no effective controls in place at the airport or ports to control the import of species and EPA has limited capacity to enforce such controls. Though there is a ban on the removal or export of Socotra flora and fauna, there are many reported incidences of smuggling out of rare endemic species, including Dragon's blood tree, affecting the biodiversity of Socotra Island. Due to ongoing regional instabilities, the coordination between major stakeholders regarding biodiversity conservation issues and decisions may also be affected. The current sociopolitical and post-Covid economic trends are also negatively impacting the site's capacity to provide sustainable economic growth for its people. The local government has limited financial resources and limited protective capacity to enforce local protective legislation. Thus, the population of Dragon's blood tree is thinning from forests to woodlands, shrublands, and eventually grasslands, with individual sparse trees.

Conservation of vulnerable and endangered endemic species depends on long-term strategies, coordination between different agencies, committed international funding, and involvement of the indigenous population with realistic developmental models; otherwise, money is being wasted with no changes on the ground, affecting the biodiversity more than before. A sustainable financing strategy also needs to be formulated to ensure necessary human and financial resources for the long-term preservation of the endangered ecosystem. More studies, planning, and appropriate linkages need to be developed and evolved for the management of the ecosystem, its buffer zones, and Socotra Biosphere Reserve. Awareness and educational activities to the natives emphasizing the fragility of islands and extinction risks are crucial. By involving the local communities and promoting them to take the lead in conservation activities by inculcating conservation knowledge and perpetuating it to the future generation, the inevitable negative impacts on biodiversity and livelihoods could be countered with improved ecosystem resilience. The local legislative laws for conservation need to be strengthened, maintaining a delicate balance among biodiversity preservation, sustainable trade, tourism, and infrastructure development. The careful implementation of these strategies is likely to positively impact the future of the endemic species in the Socotra archipelago.

3. Recommendations for the preservation of the Dragon's blood tree population

As the vulnerable biosphere reserve faces new challenges, we propose the following recommendations to protect and conserve the vulnerable Dragon's blood tree population and Socotra's unique archipelago biodiversity [5, 20, 21, 49–51]:

1. Propagating and protecting Dragon's blood tree by:

    - Ecological restoration of Dragon's blood tree species in a protected environment, including local specialized nurseries for providing scientific expertise, technical assistance, and funding to the local inhabitants for the care of the saplings. This will help in the ecological restoration of native Dragon's blood tree communities, even if it takes many years for vulnerable trees to grow to maturity (exceptionally slow growing, around 2.65 cm over a five-year old period).

    - Developing concept for a certified sustainable community forestry system, establishing a forest nursery to produce tree seedlings from local tree species population.

    - Encouraging natives for setting good practice example even at the smallest scale, by planting and nurturing a single sapling of Dragon's tree in their garden.

    - Protecting existing Dragon's blood tree and seeds from being used as fodder for livestock, firewood, and domestic use like house construction.

    - Protecting aging Dragon blood trees from natural calamities with strategies focusing in-situ conservation.

- Promoting sustainable land management practices to enable Dragon's blood tree regeneration.
- Protecting the existing Dragon's blood trees from human interference including unplanned infra-structural development/roads to mountain.
- Promoting use of UAVs (drones) in collecting crucial data and creating a drone inventory for evaluating the conservation status and threat assessment of Dragon blood tree and other vulnerable species in the Socotra ecosystem.

2. Preserving Dragon tree habitat fragmentation and woodland degradation from overgrazing by feral and domesticated goats and other cattle.
3. Promoting sustainable harvesting of Dragon's tree blood.
4. Taking initiative for curbing soil erosion, increasing ecosystem resilience, and addressing threats from unsustainable resources to the ecosystem.
5. Protecting and insulating the native ecosystem from invasive alien species (IAS) by biodiversity monitoring system, with emphasis on an early warning system with appropriate checks for invasive alien species.
6. Strengthening of the sea port and airport biological monitoring to avoid smuggling out of endangered Dragon's blood tree species.
7. Strict enforcement of the Socotra archipelago's protected area regulations and developing buffer zones in a complementary manner in Conservation Zoning Plan (CZP).
8. Extending the boundaries of the strictly protected Skund nature sanctuary (where no (road) infrastructure is allowed) to include areas of Dragon's blood tree population.
9. Enactment of local legislative laws and governance including Environmental Impact Assessment (EIA) and project approval prior to further infra-structural development and road construction.
10. Maintaining a registry with proper data compilation, processing, and management, with regular monitoring of Dragon's blood tree population and distribution by regional centers.
11. Formulating and improvising the existing master plan for sustainable infra-structural developmental activities with controlled sustainable development plans for preserving the island's Dragon's blood tree population.
12. Increasing environmental awareness in the local population and strengthening local conservation efforts by employing adequate workforce.
13. Educating, involving, and empowering the local population about the need for protecting the vulnerable Dragon's blood tree species and Socotran heritage.

14. Developing models that can assess and help in improving the chances of survival of endangered Dragon's blood tree species.
15. The development of nature-based tourism development program (Ecotourism).
16. Strong and effective collaboration among different stakeholders (from individuals to institutions).
17. Strengthening local legislative laws for conservation, aided by a strong political will to protect, preserve, and promulgate the archipelago's unique species.
18. Lastly, a sustainable financial strategy needs to be formulated to ensure long-term preservation of this flagship species.

4. Conclusion

Socotra's biodiversity remained resilient for centuries; however, the last two decades have threatened the well-preserved ecosystem including the vulnerable endemic Dragon's tree population. The strategic, result-oriented biodiversity preservation approach along with consideration of the proposed recommendations will not only help in protecting the Socotra's unique biodiversity from present and future challenges but also serve as a benchmark for biodiversity conservation around the globe.

Acknowledgements

I would like to express my deepest gratitude to Divine mother Maa MKM for all the blessings and my dear mother Mrs. Vidya Saraf for profound belief in my abilities and unconditional support. I would like to extend my sincere thanks to my dear friend Mr. Hassan Abd Elfatah Hassan Ismail, Ms. Christina and Mr. Shailesh for their unwavering help and unparalleled support. I also would like to thank dear DSS for unending inspiration and invaluable patience. Finally, I am very thankful and grateful to the staff of Summerland hotel and wonderful people of Socotra for their warm hospitality and great support during my stay in Socotra.

Conflict of interest

The author declares no conflict of interest.

Video materials

"Video materials referenced in this chapter are available at: {https://bit.ly/3Dx92h6}"

Author details

Sanjay Saraf
NMC Specialty Hospital, Dubai, United Arab Emirates (U.A.E)

*Address all correspondence to: drsaraf@hotmail.com

References

[1] Miller AG, Morris M. Ethnoflora of the Socotra Archipelago. Royal Botanic Garden: Edinburgh; 2004. p. 759

[2] Environment Protection Authority, Republic of Yemen (n.d.) Yemen Socotra Archipelago

[3] Brown G, Mies BA. Vegetation ecology of Socotra. In: Plant and Vegetation 7. 1st ed. Dordrecht, The Netherlands; Heidelberg, Germany; New York, NY, USA; London, UK: Springer; 2012. p. 379

[4] De Sanctis M, Adeeb A, Farcomeni A, Patriarca C, Saed A, Attorre F. Classification and distribution patterns of plant communities on Socotra Island, Yemen. Applied Vegetation Science. 2013;**16**:148-165

[5] Maděra P, Volařík D, Patočka Z, Kalivodová H, Divín J, Rejžek M, et al. Sustainable land use management needed to conserve the Dragon's blood tree of Socotra Island, a vulnerable endemic umbrella species. Sustainability. 2019;**11**:3557

[6] Available from: https://www.itis.gov/servlet/SingleRpt/SingleRpt?search_topic=TSN&search_value=505865#null

[7] Adolt R, Palvis J. Age structure and growth of Dracaena cinnabari populations on Socotra. Trees. 2004;**18**:43-53

[8] Axelrod DI. Evolution and biogeography of madrean-tethyan sclerophyll vegetation. In: Annals of the Missouri Botanical Garden. St. Louis, Missouri, US: Missouri Botanical Garden Press; 1975. pp. 280-334

[9] Mies BA, Beyhl FE. The Vegetation Ecology of Soqotra, Soqotra – Proceedings of the First International Symposium on Soqotra Island. Present & Future. New York: United Nations Publications; 1996. pp. 35-82

[10] Attorre F, Francesconi F, Taleb N, Scholte P, Saed A, Alfo M, et al. Will dragonblood survive the next period of climate change? Current and future potential distribution of Dracaena cinnabari (Socotra, Yemen). Biological Conservation. 2007;**138**(3-4):430-439

[11] Habrová H, Čermák Z, Pavliš J. Dragon's blood tree – Threatened by overmaturity, not by extinction: Dynamics of a Dracaena cinnabari woodland in the mountains of Soqotra. Biological Conservation. 2009;**142**(4):772-778

[12] Petroncini S. Survey and monitoring of Dracaena cinnabari Balf. Fil. in Soqotra Island (Yemen). Florence: Department of Plant Biology-Laboratory, University of Florence; 2001

[13] Mies B. Flora und Vegetationsokologie der Insel Soqotra. Essener Okologische Schriften 15. Essen, Germany: Westarp Wissenschaften; 2001

[14] Hubálková I. Prediction of Dragon's blood tree (Dracaena cinnabari Balf.) stand sample density on Soqotra Island. Journal of Landscape Ecology. 2011;**4**:5-17

[15] Wellsted JR. Memoir on the island of Socotra. The Journal of the Royal Geographical Society of London. 1835;**5**:129-229

[16] Balfour IB. Botany of Socotra. Natural History Museum Library: London, UK; 1888

[17] Král K, Pavliš J. The first detailed land cover map of Socotra Island by Landsat/ETM+ data. International Journal of Remote Sensing. 2006;**27**:3239-3250

[18] Van Damme K, Banfield L. Past and present human impacts on the biodiversity of Socotra Island (Yemen): Implications for future conservation. Zoology in the Middle East. 2011;**54**:31-88

[19] Pettorelli N, Schulte to Bühne H, Tulloch A, Dubois G, Macinnis-Ng C, Queirós AM, et al. Satellite remote sensing of ecosystem functions: Opportunities, challenges and way forward. Remote Sensing in Ecology and Conservation. 2018;**4**:71-93

[20] Al-Okaishi A. Local management system of Dragon's blood tree (Dracaena cinnabari Balf. F.) resin in Firmihin Forest, Socotra Island, Yemen. Forests. 2020;**11**:389

[21] Vahalík P, Van Damme K, Nétek R, Habrová H, Tulková J, Lengálová K, et al. UAV inventory of the last remaining dragon tree Forest on earth. Forests. 2023;**14**:766

[22] IUCN. The IUCN Red List of Threatened Species 2004. Available from: https://dx.doi.org/10.2305/IUCN.UK.2004.RLTS.T30428A9548491.en. [Accessed: 2023-08-07]

[23] Al-Okaishi A. Exploring the historical distribution of Dracaena cinnabari using ethnobotanical knowledge on Socotra Island, Yemen. Journal of Ethnobiology and Ethnomedicine. 2021;**17**:22

[24] Breasted JH. The Periplus of the Erythraean Sea: travel and trade in the Indian Ocean by a merchant of the first century, translated from the Greek and annotated by Wilfred H. Schoff, AM, Secretary of the Commercial Museum, Green, and Company: Philadelphia. London. 1912

[25] IBIDIS Press. Dioscorides. In: De Materia Medica. PO Box 81169 Parkhurst Johannesburg South Africa 2120. Ibidis Press; 2000. p. 799

[26] Ibn Sallam, Abū-Ubayd al-Qāsim. al-Gharīb al-mu-annaf, Safwān 'Adnān Dāwūdī.; dar al-fayha'a: Beirut; 2005

[27] Ad-Dīnawarī A, H. In: Lew B, editor. Book of Plants. Wiesbaden, Germany: Franz Steiner Verlag; 1974. pp. 474

[28] Manzur I. Muhammad ibn Mukarram (1290). Lisān al-ʿArab. Beirut: Dar Sader; 1990

[29] Al-Firuzabadi, Abū al-Tāhir Majīd al-Dīn (1410). Al-Qamus al-Muhit, supervised by: Muhammad Naim al-Irqsusi.; Mu'asasat al-Risalah: Beirut; 2005

[30] Casson L. The Periplus Maris Erythraei: Text with Introduction, Translation, and Commentary. Princeton, NJ, USA: Princeton University Press; 1989

[31] Cheung CPS, DeVantier LM. Socotra - A Natural History of the Islands and their People. K. Van Damme (Science Editor). Hong Kong: Odyssey Books and Guides, Airphoto International Ltd.; 2006. p. 408

[32] Damme K, editor. Hong Kong, China: Airphoto International Ltd.; 2006; 408 p

[33] Rossi D, Bruni R, Bianchi N, Chiarabelli C, Gambari R, Medici A, et al. Evaluationofthemutagenic,antimutagenic and antiproliferative potential of Croton lechleri (Muell. Arg.) latex. Phytomedicine. 2003;**10**:139-144

[34] Lopes MIL, Saffi J, Echeverrigaray S, Henriques JAP, Salvador M. Mutagenic and antioxidant activities of Croton lechleri sap in biological systems. Journal of Ethnopharmacology. 2004;**95**:437-445

[35] Gonzalez GF, Valerio LG. Medicinal plants from Peru: A review of plants as potential agents against cancer. Anti-Cancer Agents in Medicinal Chemistry. 2006;**6**:429-444

[36] Gupta D, Bleakley B, Gupta RK. Dragon's blood: Botany, chemistry and therapeutic uses. Journal of Ethnopharmacology. 2008;**115**:361-380

[37] Scholte P, De Geest P. The climate of Socotra Island (Yemen): A first-time assessment of the timing of the monsoon wind reversal and its influence on precipitation and vegetation patterns. Journal of Arid Environments. 2010;**74**:1507-1515

[38] Rejžek M, Svátek M, Šebesta J, Adolt R, Maděra P, Matula R. Loss of a single tree species will lead to an overall decline in plant diversity: Effect of Dracaena cinnabari Balf. f. on the vegetation of Socotra Island. Biological Conservation. 2016;**196**:165-172

[39] Batelka J. Socotra archipelago—A lifeboat in the sea of changes: Advancement in Socotran insect biodiversity survey. Acta Entomologica Musei Nationalis Pragae. 2012;**52** (Suppl. 2):1-26

[40] Nadezhdina N, Al-Okaishi A, Madera P. Sap flow measurements in a Socotra Dragon's blood tree (Dracaena cinnabari) in its area of origin. Tropical Plant Biology. 2018;**11**:107-118

[41] Hildebrandt A, Eltahir EAB. Forest on the edge: Seasonal cloud forest in Oman creates its own ecological niche. Geophysical Research Letters. 2006;**33**:1-4

[42] García C, Vasconcelos R. The beauty and the beast: Endemic mutualistic interactions promote community-based conservation on Socotra Island (Yemen). Journal for Nature Conservation. 2017;**35**:20-23

[43] Habrová H, Pavliš J. Dynamic response of woody vegetation on fencing protection in semi-arid areas; case study: Pilot exclosure on the Firmihin plateau, Socotra Island. Saudi Journal of Biological Sciences. 2017;**24**:338-346

[44] Fleitmann D, Burns SJ, Mangini A, Mudelsee M, Kramers J, Villa I, et al. Holocene ITCZ and Indian monsoon dynamics recorded in stalagmites from Oman and Yemen (Socotra). Quaternary Science Reviews. 2007;**26**:170-188

[45] J.R. Malcolm et al. Global Warming and Extinctions of Endemic Species from Biodiversity Hotspots Conservation Biology. 2006

[46] Midgley GF, et al. Assessing the vulnerability of species richness to anthropogenic climate change in a biodiversity

[47] Kalivodová H, Culek M, Čermák M, Maděra P, Habrová H. Potential importance of Socotra dragon's blood tree cloud forests and woodlands for capturing horizontal precipitation. Rendiconti Lincei. 2020;**31**(3):607-621

[48] Van Rensburg JJ, Hopper K. Incense and imagery: Mapping agricultural and water management systems on the island of Socotra, Yemen. Proceedings of the Seminar for Arabian Studies. 2017;**47**:129-138

[49] Maděra P, Habrová H, Šenfeldr M, Hubálková I, Lvončík S, Ehrenbergerová L,

et al. Growth dynamics of endemic Dracaena cinnabari of Socotra Island suggest essential elements for a conservation strategy. Biologia. 2019;**74**:339-349

[50] Attorre F, Taleb N, De Sanctis M, Farcomeni A, Guillet A, Vitale M. Developing conservation strategies for endemic tree species when faced with time and data constraints: Boswellia spp. on Socotra (Yemen). Biodiversity and Conservation. 2011;**20**:1483-1499

[51] Saraf S. Preserving the perishing endangered natural biodiversity of Socotra Island. Open Journal of Ecology. 2021;**11**:148-162

Chapter 6

The Economics of Endangered Species in Hawaii during the COVID-19 Pandemic

Patricia Yu

Abstract

The COVID-19 pandemic has significantly impacted Hawaii's vulnerable ecosystem of endangered species. Despite some scholars suggesting that the pandemic has offered a silver lining by allowing the environment to regenerate and create a safer habitat for these species, the economic impacts of the pandemic cannot be ignored. This paper aims to explore the economics of endangered species in Hawaii during the COVID-19 pandemic. The reduction in tourism has led to improvements in air quality and traffic congestion, as well as increased conservation efforts. However, the decrease in tourism has also had a negative impact on the economy, as tourism is a significant contributor to Hawaii's economy. This paper analyzes the economic trade-offs between conservation efforts and economic growth and explore potential solutions to ensure the long-term sustainability of Hawaii's endangered species and economy.

Keywords: COVID-19 pandemic, endangered species, economy, tourism, Hawaii

1. Introduction

Hawaii is home to a unique and diverse range of plant and animal species, many of which are endangered. A report from the US Congress' Office of Technology Assessment [1] stated that because of being a hub for trade, tourism, and military activities: [T]he Hawaiian Islands represent the worst-case example of the Nation's NIS (non-indigenous species) problem. No other area in the United States receives as many new species annually, nor has as great a proportion of NIS established in the wild. At the same time, Hawaii, the Nation's so-called extinction capital, has the greatest concentration of threatened and endangered species in the United States and

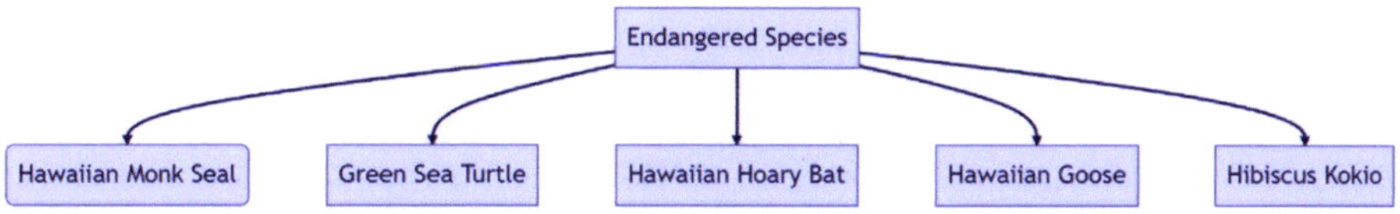

Figure 1.
A sample of endangered species in Hawaii.

the greatest number of extinct species as well (p. 234). See **Figure 1** for a sample of the endangered species in Hawaii.

The COVID-19 pandemic has had a substantial impact on Hawaii's vulnerable ecosystem of endangered species, despite some scholars suggesting that the pandemic has offered a silver lining by allowing the environment to regenerate and create a safer habitat for these species.

Oleson et al. [2] provide a comprehensive overview of the impact of COVID-19 on the economy and environment of Hawaii. It discusses both the negative impacts, such as the loss of jobs and revenue in the tourism industry and the positive impacts, such as the reduction in air pollution and the potential for increased sustainability efforts. They argue that the pandemic provides an opportunity to rethink and transform Hawaii's economy and society in a more sustainable direction. One of the noticeable impacts is the reduction in tourism, which has also led to a significant reduction in traffic congestion in Hawaii. The lack of tourists and reduced commuting have led to less traffic on the roads, resulting in shorter travel times, reduced emissions, and improved safety. The COVID-19 pandemic has also led to some levels of increased conservation efforts in Hawaii. Several organizations, including the Hawaii Nature Center, have implemented new programs to engage residents in conservation efforts, including beach cleanups, native plant restoration, and wildlife monitoring. The Nature Conservancy [3] provides an overview of the impact of COVID-19 on conservation efforts in Hawaii. It highlights both the challenges and opportunities presented by the pandemic, including disruptions to fieldwork and the need to find new ways of engaging with local communities. The resource also emphasizes the importance of continuing conservation efforts during this time of crisis. Iniguez and D'Antonio [4] focuses on the impact of COVID-19 on ambient noise levels in Hawaii, which decreased significantly due to the reduction in human activity during the pandemic. The authors argue that this reduction in noise pollution could have positive impacts on local wildlife, particularly marine mammals that rely on sound for communication and navigation.

The decrease in tourism seems to have a positive impact on Hawaii's wildlife. With fewer people visiting popular tourist destinations, such as beaches and parks, wildlife has been able to thrive in their natural habitats without the disturbances caused by human activities. Several endangered species, including the Hawaiian monk seal and green sea turtle, have been observed nesting and laying eggs on beaches that are normally crowded with tourists. This paper comprises several sections that delve into the economics, positive externalities, conservation efforts, and technologies associated with endangered species in Hawaii. The first section provides a general description of the key economic models measuring the values of endangered species, while the second section explores the positive externalities of endangered species. Subsequently, this paper discusses Hawaii's ongoing efforts to protect these endangered species, followed by an analysis of the technologies utilized in their conservation. This paper concludes with a section on potential solutions to Hawaii's endangered species.

2. Economics of endangered species in Hawaii

Many literatures explore the use of an economic model to value the endangered species. Diamond and Hausman [5] uses contingent valuation methods to estimate the value of non-market goods, including endangered species. They argue that while contingent valuation has some limitations, it is still a useful tool for policymakers in estimating the economic value of environmental resources. Hanley and Spash [6] uses the cost-benefit analysis (CBA) as a tool for decision-making in environmental policy, including policies related to endangered species. They argue that while CBA has limitations, it is still a valuable tool for policymakers in determining the most efficient use of resources. Pagiola et al. [7] provides a comprehensive overview of methods for assessing the economic value of ecosystem conservation, including the value of endangered species. They argue that while assigning economic values to ecosystem services can be challenging, it is essential for decision-making in environmental policy. This report also provides case studies and examples of successful economic valuation of ecosystem services in practice.

CBA could be used to compare the costs of captive breeding programs for endangered birds in Hawaii versus the benefits of ecotourism revenue generated by the presence of these birds in the wild. Similarly, CBA could be used to compare the costs and benefits of habitat restoration in Hawaii versus the costs and benefits of establishing protected areas. However, it is important to note that CBA has limitations, and the economic value of ecosystem services and endangered species may be difficult to quantify in some cases. Nonetheless, CBA can be a valuable tool for decision-making in environmental policy and can help to ensure that conservation resources are used in the most efficient and effective manner possible.

The presence of endangered species in Hawaii has a significant impact on local communities. For example, the Hawaiian monk seal is a species that is only found in Hawaii, and its survival is crucial for the Hawaiian culture. One of the primary methods of monitoring the Hawaiian Monk seal is through fieldwork. However, due to the pandemic, many of the fieldwork activities had to be suspended or reduced. This reduction in monitoring has made it difficult to track the population of the Hawaiian Monk seal accurately. The lack of monitoring could also make it easier for poachers to hunt these seals. The pandemic has also resulted in a reduction in funding and resources for conservation efforts for the Hawaiian Monk seal. Many conservation programs rely on volunteers and field workers, but the pandemic has made it difficult for these workers to carry out their duties. The reduced funding has also limited the number of conservation efforts that can be undertaken, making it difficult to protect the seals from the threats they face. CVM can be used to estimate the economic value of the Hawaiian monk seal. This would involve surveying individuals in Hawaii and other locations to determine their willingness-to-pay for the conservation of the Hawaiian monk seal. The survey would provide participants with information about the Hawaiian monk seal and its status as a critically endangered species. Participants would then be asked how much they would be willing to pay to support conservation efforts for the Hawaiian monk seal. By using CVM, the economic value of the Hawaiian monk seal and the factors that influence individuals' willingness-to-pay for its conservation can be revealed. The results of the survey can be used to inform conservation decisions related to the Hawaiian monk seal. For example, if the estimated economic value of the Hawaiian monk seal is high, policymakers may be more likely to allocate resources to its conservation. However, it is important to note that CVM also has limitations, and the results may be influenced by factors such as survey design and participant biases.

Hawaii has experienced a significant loss of species due to various human activities, such as habitat destruction, invasive species, and climate change. The suite of challenges facing island endemic species, including invasive predators and competitors, habitat degradation, and the loss of mutually beneficial species, alongside economic and social challenges lead to a complex decision environment [8]. The loss of species can have a profound impact on the tourism industry in Hawaii. The declining tourism industry would have a significant impact on the state's economy. With fewer unique species to showcase, Hawaii may struggle to differentiate itself from other tourist destinations, ultimately leading to a reduction in tourism revenue. Even with many diversifying efforts from the Hawaiian governments, the tourism industry remains a critical component of Hawaii's economy, providing jobs and revenue for many locals. The loss of species can negatively impact the local economy, especially for those involved in the tourism industry. The loss of species can also have a significant impact on the environment of Hawaii. Many species play important roles in the ecosystem, such as pollination and seed dispersal, and their loss can disrupt the delicate balance of the environment. The loss of species can also have ripple effects on other species, leading to a decline in biodiversity, which can ultimately impact the tourism industry in Hawaii. Many of the endangered species in Hawaii are deeply rooted in Hawaiian culture and history. For example, the Hawaiian green sea turtle, known as "honu", is a revered animal in Hawaiian culture and is often seen as a symbol of good luck and longevity. Protecting these species is not only important for their ecological value but also for their cultural significance.

3. Positive externalities of endangered species

There are several key literatures on the positive externalities of endangered species. Brander and Taylor [9] presents a theoretical model of renewable resource use, specifically focusing on Easter Island as a case study. They argue that the presence of endangered species, such as seabirds, provided positive externalities in the form of ecosystem services and contributed to the sustainability of the island's economy. This paper highlights the importance of considering positive externalities in the valuation of natural resources, particularly in the context of renewable resource use. Ferraro [10] presents a method for targeting conservation investments in heterogeneous landscapes, considering the positive externalities of endangered species. The author uses a distance function approach to estimate the impact of conservation investments on both economic outcomes and the distribution of ecosystem services, including those provided by endangered species. This paper highlights the importance of considering positive externalities in the design of conservation programs and the potential for targeted conservation investments to generate positive economic outcomes. Ribaudo et al. [11] presents the results of a survey of economists' opinions on the environmental benefits of conservation practices, including the positive externalities of endangered species. They find that economists generally agree that endangered species provide positive externalities in the form of ecosystem services and that these services should be considered in the valuation of conservation practices. This paper highlights the importance of considering positive externalities in the design of conservation policies and the potential for market-based mechanisms to encourage the provision of ecosystem services.

The first kind of positive externality of endangered species in Hawaii is towards the ecotourism industry. People from all over the world travel to Hawaii to see the unique and rare species that exist only on the islands. This ecotourism industry provides jobs and revenue for the local economy, as well as educational opportunities for tourists. Ecotourism aims to minimize the negative impact of tourism while maximizing its positive impact on the environment and local communities. By encouraging visitors to appreciate the natural beauty of Hawaii, ecotourism helps to protect its unique environment and ecosystem. It provides economic benefits to local communities by creating jobs and supporting small businesses. It also promotes cultural exchange and understanding between tourists and residents as well as provides visitors with an opportunity to learn about Hawaii's unique environment, culture, and history. Nonetheless, ecotourism has its own negative impact on our environment. Ha and Ha [12] examines the environmental impacts of ecotourism in Hawaii, focusing on the impacts of visitor behavior and the infrastructure needed to support ecotourism activities. They found that while ecotourism can have positive environmental impacts, such as promoting conservation and sustainable practices, it can also have negative impacts, such as damage to natural habitats and increased pollution. This article concludes with recommendations for sustainable ecotourism practices in Hawaii.

The second kind of positive externality of endangered species in Hawaii is reflected in the genetic diversity. Endangered species in Hawaii represent a unique genetic pool that can provide valuable genetic diversity. This diversity can be used to improve the genetic health of other populations and can be used to breed hybrids that may be more resilient to changing environmental conditions. This genetic diversity can also be used for medical research, leading to new treatments and cures. One of the most noticeable impacts of the COVID-19 pandemic on genetic diversity in Hawaii has been the decrease in human activity. With travel restrictions and social distancing measures in place, there has been a decrease in the number of people visiting Hawaii. This decrease in human activity has allowed some species to thrive and increase their genetic diversity. For example, the green sea turtle has seen an increase in nesting and hatching success due to reduced human activity on beaches. This increase in nesting success may lead to an increase in genetic diversity, as successful reproduction is crucial for maintaining genetic diversity in populations. While the decrease in human activity has had some positive effects on genetic diversity, there are also potential long-term effects on biodiversity. The COVID-19 pandemic has highlighted the importance of protecting biodiversity and the interconnectedness of ecosystems. Reduced human activity may have unintended consequences on ecosystems, including changes in habitat availability and food availability for species. Kitayama et al. [13] examine the impact of human activities, such as deforestation and invasive species, on the nutrient cycling and soil properties of Hawaiian montane rainforests. They found that these activities can have a significant impact on the health and biodiversity of the ecosystem. This article highlights the importance of preserving intact ecosystems and preventing further degradation to maintain biodiversity.

The third kind of positive externality of endangered species in Hawaii is the ecosystem service. Endangered species in Hawaii provide ecosystem services that are often taken for granted. For example, the Hawaiian hoary bat is a key pollinator for many of Hawaii's native plant species. These plants, in turn, provide habitat and food for other species, such as the endangered Hawaiian monk seal. Without these ecosystem services, the delicate balance of Hawaii's unique ecosystem could be disrupted. The silver lining of the COVID-19 pandemic is reflected in this ecosystem service's relative betterment. The reduction in air and water pollution from reduced tourism and transportation has improved the air and water quality in Hawaii, leading to an increase in ecosystem services such as clean air and water. Additionally, the decrease in human activity has allowed natural habitats to recover, increasing biodiversity and ecosystem functions such as carbon sequestration and soil formation. While the decrease in human activity has had some positive effects on ecosystem services, there are also potential long-term effects on the environment and society. The COVID-19 pandemic has highlighted the importance of protecting and enhancing ecosystem services, which are essential for human well-being and sustainable development. However, the pandemic has also highlighted the vulnerabilities of ecosystems and the need for increased resilience and adaptive management. The decrease in tourism and associated economic activity has affected the ability of local communities to maintain their livelihoods and provide for their families, potentially leading to social and economic challenges. Carrington [14] discusses the link between the destruction of natural habitats and the emergence of zoonotic diseases like COVID-19. It emphasizes the importance of protecting ecosystems and biodiversity to prevent future pandemics. This article highlights the need for policies and legislation to address the root causes of pandemics and promote a green recovery.

4. Efforts to protect endangered species in Hawaii

The state of Hawaii has taken significant steps to protect its endangered species. For example, the state has established several conservation programs to protect endangered species and their habitats. These programs include habitat restoration, captive breeding, and reintroduction programs. The state has also enacted laws and regulations to protect endangered species. The Endangered Species Act (ESA) of 1973 is a federal law that provides protection for endangered and threatened species and their habitats. Fischman [15] provides an overview of the ESA, its history, and its effectiveness in protecting endangered species. The author argues that the ESA is a powerful tool for protecting endangered species and that its successes can be attributed to its strong legal framework and the support of dedicated professionals and conservation organizations. This paper highlights the importance of political will and public support in the efforts to protect endangered species. Wilcove and Master [16] provides an assessment of the number of endangered species in the United States and the effectiveness of conservation efforts in protecting them. They argue that the ESA has been successful in preventing the extinction of many endangered species, but that additional efforts are needed to recover populations and improve the status of threatened species. This paper highlights the need for continued monitoring and adaptive management in the efforts to protect endangered species. Oates and Myers [17] presents the 1985 International Union for Conservation of Nature (IUCN) Red List of Threatened Animals, which was a significant milestone in the efforts to protect endangered species. The Red List provides a comprehensive assessment of the conservation status of species around the world and has since become a widely recognized tool for identifying and prioritizing conservation efforts. This paper highlights the importance of international collaboration and data sharing in the efforts to protect endangered species.

Hawaii has also enacted state laws to protect endangered species, such as the Hawaii Endangered Species Act. The state has also collaborated with local communities, non-profit organizations, and federal agencies to protect endangered species. These collaborations have been crucial in the success of conservation efforts in Hawaii. For example, the Hawaii Department of Land and Natural Resources has increased their efforts to protect native forest birds, while the University of Hawaii has been studying the effects of reduced human activity on marine mammals. The success of conservation efforts in protecting endangered species in Hawaii during the COVID-19 pandemic has been significant. VanderWerf and Young [18] examines the impacts of COVID-19 on endangered bird conservation efforts in Hawaii. They found that while the pandemic has had some negative impacts on fieldwork, such as reduced funding and volunteer participation, it has also provided some unexpected benefits, such as reduced disturbance from tourists and increased community involvement in conservation efforts. This article provides insights into how unexpected events like pandemics can impact conservation efforts and highlights the importance of community involvement in conservation. Additionally, conservation efforts to remove invasive species and restore native habitats have been successful. The removal of feral cats and rats from the island of Lehua has allowed for the restoration of native bird habitats, resulting in the successful reintroduction of endangered seabirds. While the decrease in human activity has had some positive effects on conservation efforts, there are also potential long-term effects on the environment and society. The COVID-19 pandemic has highlighted the importance of protecting and enhancing conservation efforts, which are essential for the preservation of endangered species and the long-term health of ecosystems. However, the pandemic has also highlighted the vulnerabilities of ecosystems and the need for increased resilience and adaptive management. The decrease in tourism and associated economic activity has affected the ability of local communities to maintain their livelihoods and provide for their families, potentially leading to social and economic challenges.

The Hawaiian culture has a deep connection between people's livelihood to the land and its natural resources, reflected in the concept of "kuleana", which refers to one's responsibility and accountability to care for the environment. Leopold [19] explores the concept of kuleana and argues that incorporating this value into conservation efforts in Hawaii can increase community involvement and support for conservation, as it emphasizes the responsibility of individuals and communities to care for the land and its resources. This article provides insights into the cultural values that can inform and shape conservation efforts in Hawaii and highlights the potential for indigenous values to increase community involvement in conservation.

5. Technology of protecting endangered species in Hawaii

There are several literatures on using technologies to protect endangered species. Shahriar et al. [20] provides an overview of the different technologies that are being used to protect endangered species. They describe the use of GPS tracking, remote sensing, and drones in monitoring and conserving species, as well as the use of genetic analysis and biotechnology in understanding and preserving genetic diversity. This paper highlights the potential of these technologies in improving the efficiency and effectiveness of conservation efforts. Duran et al. [21] provides a comprehensive review of the use of biotechnology in protecting endangered species. They describe the use of techniques such as artificial insemination, embryo transfer, and cryopreservation in preserving genetic diversity and enhancing reproductive success. This paper highlights the potential of biotechnology in improving the genetic health and resilience of endangered species. Gober and Kumar [22] focuses on the use of remote sensing technology in monitoring and managing endangered species. They describe the use of satellite imagery, unmanned aerial vehicles (UAVs), and ground-based sensors in detecting and mapping habitats, monitoring populations, and assessing the impacts of environmental changes. This paper highlights the potential of remote sensing technology in improving the efficiency and accuracy of conservation efforts, and the need for continued research and development in this area.

As the world faces an unprecedented rate of biodiversity loss, technology has become an increasingly important tool in protecting endangered species. In Hawaii,

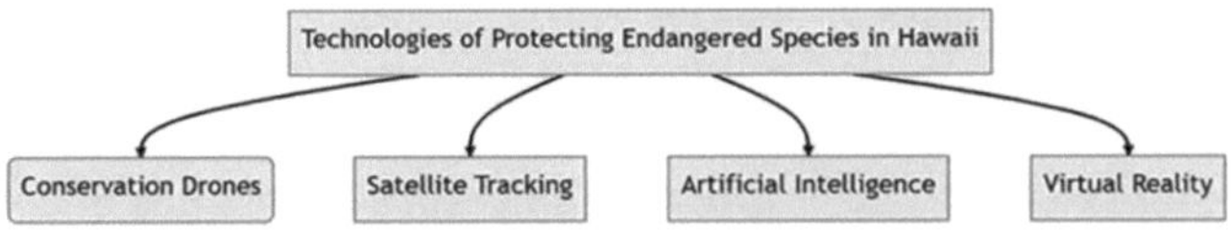

Figure 2.
Technologies of protecting endangered species in Hawaii.

new technologies are being developed and implemented to address the threats facing endangered species. See **Figure 2** for a glimpse of the technologies used in Hawaii:

The first technology in **Figure 2** is the conservation drones, which have become a useful tool in the protection of endangered species in Hawaii. These unmanned aerial vehicles (UAVs) can be used to monitor and survey wildlife populations, track migration patterns, and detect illegal activities such as poaching and logging. Drones have also proven to be valuable tools in the discovery, inventory, and mapping of rare cliff plants; and due to the difficulty of the on-the-ground survey of cliffs, drones have been deployed across a range of environments leading to the discovery of unknown populations [23]. The other technology is satellite tracking, which has become a popular tool in the protection of endangered species in Hawaii. By attaching satellite tags to animals, researchers can track their movements, migration patterns, and behavior. This information can be used to identify critical habitats, monitor population trends, and develop effective conservation strategies. In Hawaii, satellite tracking has been used to monitor the movements of endangered sea turtles, monk seals, and humpback whales. Artificial intelligence (AI) has the potential to revolutionize the protection of endangered species in Hawaii. By using machine learning algorithms, AI can analyze large amounts of data, such as satellite imagery, and identify patterns that indicate the presence of endangered species or threats to their habitats. This can help researchers and conservationists identify areas in need of protection and develop effective conservation strategies. In Hawaii, AI has been used to identify nesting sites of the endangered Hawaiian petrel and to monitor the spread of invasive species. AI can also be applied to large groups of animals (e.g., ant colonies and beehives), or implemented for monitoring animals that are often difficult to sample, namely cryptic species such as the nocturnal spotted-tailed quoll or species that cover an extensive range such as seabirds [24]. Virtual reality (VR) has emerged as a tool for raising public awareness and education about endangered species in Hawaii. By using VR technology, people can experience the habitats and behaviors of endangered species and gain a deeper understanding of the threats facing these species and the importance of their protection. In Hawaii, VR has been used to educate visitors and locals about the endangered Hawaiian monk seal and to promote conservation efforts. Putrino et al. [25] provides a review of the use of augmented and virtual reality in wildlife conservation, including its potential applications for education, research, and public outreach. They discuss the potential for virtual reality to create immersive experiences that promote empathy and understanding of environmental issues, as well as its potential for research and monitoring of endangered species. This article provides a comprehensive overview of the potential applications of virtual reality in wildlife conservation and highlights the importance of using technology to enhance conservation efforts.

6. Potential solutions to Hawaii's endangered species

There are several literatures on giving certain solutions to Hawaii's specific or general endangered species. Kelly et al. [26] focus on the impact of climate change on Hawaiian forest birds, which are highly endangered due to habitat loss and the introduction of non-native species. They argue that conservation efforts need to consider the potential impacts of climate change on the birds' habitats, as well as the need for landscape-level management that integrates both natural and cultural values. This paper provides insights into the challenges and opportunities of conserving endangered species in the face of a changing climate. National Oceanic and Atmospheric Administration (NOAA) provides information on the Hawaiian monk seal and their website outlines the various efforts being undertaken to protect the species, including habitat restoration, predator control, and public education. It also highlights the importance of partnerships between government agencies, conservation organizations, and local communities in the conservation of the species [27]. Tredick et al. [28] discusses the challenges and opportunities of managing endangered species under a changing climate, using the Hawaiian hoary bat as a case study. They argue that effective conservation strategies need to integrate both short-term and long-term climate projections, as well as incorporate adaptive management approaches that allow for ongoing evaluation and adjustment of conservation measures. This paper highlights the need for innovative solutions that balance the protection of endangered species with the realities of a changing world.

With the tourism industry picking up and marching towards the post-pandemic era, the growth of tourism and associated development has again placed significant pressure on Hawaii's natural resources, threatening the long-term sustainability of both its economy and endangered species. Smith et al. [29] examines the impact of the COVID-19 pandemic on conservation science, including the disruption to fieldwork, funding, and scientific collaboration. They argue that the pandemic has highlighted the need for more resilient and adaptive conservation practices, as well as the importance of interdisciplinary approaches to conservation science. This article provides insights into the impact of the pandemic on conservation science and highlights the need for innovative and adaptive conservation strategies in the post-pandemic era.

In the face of climate change, Hawaii Island's survival is contingent upon embracing sustainability as a crucial element. Some potential solutions to ensure the long-term sustainability of Hawaii's endangered species and economy include (1) habitat restoration: the restoration of native habitats can help create safe and suitable environments for endangered species as well as opportunities for eco-tourism, which can provide revenue for local communities while protecting the environment. Habitat restoration efforts can be challenging, requiring significant resources, including funding, expertise, and time. Additionally, restoration efforts must be carefully planned and executed to avoid unintended consequences, such as the introduction of non-native species or the disruption of ecosystems. Despite the challenges, habitat restoration efforts in Hawaii have had some notable successes. For example, the restoration of the Hakalau Forest National Wildlife Refuge on the Big Island has led to the recovery of several endangered bird species. Additionally, the restoration of Kāne'ohe Bay on Oahu has resulted in the return of numerous marine species, including sea turtles and monk seals. (2) Invasive species management: invasive species pose a significant threat to Hawaii's biodiversity. One solution to manage invasive species is through effective control and eradication programs. Several methods have been used to control invasive species in Hawaii, including mechanical, chemical, and biological control methods. Mechanical methods involve physically removing invasive species, such as through hand-pulling or mowing. Chemical methods involve the use of herbicides or pesticides to control invasive species. Biological control methods involve the introduction of natural enemies of invasive species, such as predators, parasites, or diseases. Eradication programs are designed to eliminate invasive species from a particular area. These programs often require intensive efforts and significant resources. One example of an eradication program in Hawaii is the effort to eradicate the coqui frog, a highly invasive species that has spread across the islands. The program involves the use of a variety of control methods, including hand capture, habitat modification, and the use of acoustic deterrents. Prevention programs often focus on early detection and rapid response to new invasive species introductions. Hawaii has implemented several prevention programs, including the Clean, Drain, and Dry program, which aims to prevent the spread of invasive species through watercraft. Despite the challenges of controlling invasive species, there have been some notable successes in Hawaii. For example, the successful eradication of the Miconia plant from Maui has resulted in the recovery of several native bird species. Additionally, the implementation of the Little Fire Ant Eradication Program has prevented the spread of this invasive species to other islands. (3) Sustainable tourism: tourism is a significant driver of Hawaii's economy, but it can also be a source of environmental degradation. One potential solution is to shift towards sustainable tourism practices that prioritize the preservation of Hawaii's natural and cultural resources. This can include promoting eco-tourism, encouraging responsible tourism practices, and implementing sustainable development initiatives. Agriculture is an essential part of Hawaii's economy, but it can also have a significant impact on endangered species. Sustainable agriculture practices, such as crop rotation and the use of organic fertilizers, can help to reduce the impact of agriculture on endangered species. In addition, sustainable agriculture can help to promote biodiversity by preserving natural habitats and supporting the growth of native plants. (4) Local community engagement: local communities play a vital role in protecting Hawaii's endangered species and economy. One potential solution is to engage local communities in conservation efforts. This can include providing education and training on sustainable practices, involving local communities in conservation planning and decision-making, and creating economic opportunities that promote conservation. (5) Research and technology: advancements in technology can provide innovative solutions to protect Hawaii's endangered species and economy. Additionally, research into sustainable tourism practices and conservation strategies can inform policy decisions and help identify effective solutions.

7. Conclusion

Hawaiian culture has a strong connection to the land and the environment, with the concept of "malama aina" emphasizing the importance of caring for and protecting the land. Malama Aina is a central concept in Hawaiian culture, emphasizing the importance of living in harmony with the environment and protecting it for future generations. Native Hawaiians recognized this relationship in the proverb "He ali'i ka 'aina; he kauwa ke kanaka (The land is a chief; man is its servant [30]. The recognition of limited resources, the cultural ethic of conservation (malama i ka 'aina, literally to care for that which feeds), and a cultural emphasis on recognizing and documenting changes in Hawaiian ecosystems support active ecological literacy oriented to sustainability [31]. Traditional conservation practices include the concept of "ahupua'a," which is a system of land management that recognizes the interconnectedness of land, water, and people. In this system, each ahupua'a was managed to ensure the sustainability of resources such as water, fish, and plants. This system helped to prevent overexploitation and ensured the long-term health of ecosystems. Hawaii has one of the highest rates of species extinction in the world, with many of its native species facing extinction due to habitat loss, invasive species, and climate change. Protecting endangered species in Hawaii is critical to preserving the state's biodiversity and ensuring the long-term health of ecosystems. Current conservation efforts in Hawaii include the restoration of native habitats, the removal of invasive species, and the protection of endangered species through conservation programs and partnerships with local communities. These efforts align with the principles of "malama aina", emphasizing the importance of caring for the land and preserving its natural resources. Protecting endangered species in Hawaii is critical to preserving the state's biodiversity and ensuring the long-term health of ecosystems. It is essential to continue promoting and implementing sustainable conservation practices that balance economic, social, and environmental goals and preserve the unique culture and biodiversity of Hawaii for future generations.

Author details

Patricia Yu
University of Hawaii—West Oahu, Hawaii

*Address all correspondence to: pyu@hawaii.edu

References

[1] US Congress Office of Technology Assessment. Harmful Non-Indigenous Species in the United States, OTA-F-565. Washington, DC: US Government Printing Office; 1993

[2] Oleson KL, Barnes M, McPhaden MJ, Wongbusarakum S, Hurd T, Lecky J, et al. Effects of the COVID-19 pandemic on the economy and environment of Hawaii. Environmental Science & Policy. 2021;**116**:225-235. DOI: 10.1016/j.envsci.2020.11.015

[3] The Nature Conservancy. Hawaii: COVID-19 and Conservation. 2020. Available from: https://www.nature.org/en-us/about-us/where-we-work/united-states/hawaii/stories-in-hawaii/hawaii-covid19-and-conservation/

[4] Iniguez M, D'Antonio C. An unexpected and rapid reduction in ambient noise levels due to the COVID-19 pandemic in Hawai'i. Science of the Total Environment. 2020;**745**:141155. DOI: 10.1016/j.scitotenv.2020.141155

[5] Diamond PA, Hausman JA. Contingent valuation: Is some number better than no number? Journal of Economic Perspectives. 1994;**8**(4):45-64. Available from: https://www.jstor.org/stable/2138552

[6] Hanley N, Spash CL. Cost-benefit Analysis and the Environment. Edward Elgar Publishing; 1993. Available from: https://www.elgaronline.com/view/9781852785704.xml

[7] Pagiola S, von Ritter K, Bishop J. Assessing the Economic Value of Ecosystem Conservation. The World Bank; 2004. Available from: https://openknowledge.worldbank.org/handle/10986/14930

[8] Price MR, Toonen RJ. Scaling up restoration efforts in the Pacific Islands: A call for clear management objectives, targeted research to minimize uncertainty, and innovative solutions to a wicked problem. Pacific Science. 2017;**71**(4):391-399

[9] Brander JA, Taylor MS. The simple economics of Easter Island: A Ricardo-Malthus model of renewable resource use. American Economic Review. 1998;**88**(1):119-138. Available from: https://www.jstor.org/stable/116796

[10] Ferraro PJ. Targeting conservation investments in heterogeneous landscapes: A distance function approach and application to watershed management. American Journal of Agricultural Economics. 2004;**86**(4):905-918. DOI: 10.1111/j.0002-9092.2004.00641.x

[11] Ribaudo MO, Hellerstein D, Claassen R. Identifying the environmental benefits of conservation practices: A survey of economists' opinions. Ecological Economics. 2008;**65**(3):407-419. DOI: 10.1016/j.ecolecon.2007.07.008

[12] Ha TT, Ha MQ. Environmental impacts of ecotourism in Hawaii: An exploratory study. Sustainability. 2019;**11**(10):2909. DOI: 10.3390/su11102909

[13] Kitayama K, Mueller-Dombois D, Vitousek PM. Dynamics of nutrient cycling and soil properties in Hawaiian montane rainforest. Ecology. 1995;**76**(1):259-272. DOI: 10.2307/1940643

[14] Carrington D. Pandemics result from destruction of nature, say UN and WHO. The Guardian. 2020. Available

from: https://www.theguardian.com/world/2020/jun/17/pandemics-destruction-nature-un-who-legislation-trade-green-recovery

[15] Fischman RL. The power of the endangered species act. Environmental Law. 1997;**27**(4):1081-1104. Available from: https://scholarship.law.umt.edu/cgi/viewcontent.cgi?article=2233&context=faculty_lawreviews

[16] Wilcove DS, Master LL. How many endangered species are there in the United States? Frontiers in Ecology and the Environment. 2005;**3**(8):414-420 10.1890/1540-9295(2005)003[0414:HMESAT]2.0.CO;2

[17] Oates JF, Myers N. The 1985 IUCN red list of threatened animals. Conservation Biology. 1986;**1**(2):118-119. DOI: 10.1111/j.1523-1739.1987.tb00027.x

[18] VanderWerf EA, Young LC. The impacts of COVID-19 on the conservation of endangered birds in Hawai'i. Biological Conservation. 2021;**253**:108935. DOI: 10.1016/j.biocon.2020.108935

[19] Leopold AC. The Hawaiian concept of kuleana (responsibility) and its potential to increase conservation activities. Conservation Biology. 2013;**27**(3):534-538. DOI: 10.1111/cobi.12048

[20] Shahriar S, Hossain MS, Alam MF. Applications of technology in endangered species conservation. International Journal of Advanced Computer Science and Applications. 2017;**8**(4):292-302. DOI: 10.14569/IJACSA.2017.080444

[21] Duran A, González-Acuña D, Castro-Prieto J. Use of biotechnology for conservation of endangered species: A review. Journal of Veterinary Research. 2019;**63**(2):171-180. DOI: 10.2478/jvetres-2019-0023

[22] Gober P, Kumar R. The use of remote sensing technology for monitoring and managing endangered species. Journal of Environmental Management. 2019;**250**:109493. DOI: 10.1016/j.jenvman.2019.109493

[23] La Vigne H, Charron G, Rachiele-Tremblay J, et al. Collecting critically endangered cliff plants using a drone-based sampling manipulator. Scientific Reports. 2022;**12**:14827. DOI: 10.1038/s41598-022-17679-x

[24] Congdon JV, Hosseini M, Gading EF, Masousi M, Franke M, MacDonald SE. The future of artificial intelligence in monitoring animal identification, health, and behaviour. Animals. 2022;**12**(13):16. DOI: 10.3390/ani12131711

[25] Putrino N, Swords S, Brooks K, et al. Augmented and virtual reality for wildlife conservation: A review from the field. Animals. 2019;**9**(12):1114. DOI: 10.3390/ani9121114

[26] Kelly MJ, Conant S, Wilcox B. Conserving Hawaiian forest birds in the face of climate change. Biological Conservation. 2018;**227**:343-350. DOI: 10.1016/j.biocon.2018.09.002

[27] National Oceanic and Atmospheric Administration (NOAA). Hawaiian Monk Seal Recovery. 2020. Available from: https://www.fisheries.noaa.gov/species/hawaiian-monk-seal

[28] Tredick CA, Kaufman L, Kendall WL. Managing endangered species under climate change: Challenges and opportunities. Frontiers in Ecology and the Environment. 2015;**13**(1):20-28. DOI: 10.1890/140252

[29] Smith KF, Goldberg M, Rosenthal S, Carlson L, Chen J,

Chen C, et al. COVID-19's uneven toll on conservation science: The impact of the pandemic and opportunities for recovery. Biological Conservation. 2021;**257**:109145. DOI: 10.1016/j.biocon.2021.109145

[30] Pukui MK. 'Olelo no'eau: Hawaiian Proverbs and Poetical Sayings. Honolulu: Bishop Museum Press; 1983

[31] Chinn P, Mālama I. Ka 'Āina, sustainability: Learning from Hawai'i's displaced place and culture-based science standard. Cultural Studies of Science Education. 2011;**6**(1):223-233. DOI: 10.1007/s11422-011-9312-z